Advance Praise for
Stress-Proof

"Extensively researched and comprehensive, *Stress-Proof* is filled with fascinating strategies for preventing chronic stress. Its advice is powerful and yet simple to implement and promises tremendous benefits for both mental and physical well-being."
—Dan Buettner, National Geographic fellow
and *New York Times* bestselling author of *The Blue Zones*

"*Stress-Proof* is a rigorously researched guide that presents cutting-edge strategies for improving resilience, mental performance, and focus. Highly recommended."
—Scott Barry Kaufman, PhD; psychologist
and coauthor of *Wired to Create*

"Helpful and practical. Applying this book to your life will make it better."
—Kamal Ravikant, bestselling author of
Love Yourself Like Your Life Depends On It

"Mithu Storoni explains the neurobiology of stress and provides wise and accessible advice on enabling happy resilience. We learn why many practical steps can help us thrive in our stressful lives."
—Dame Sandra Dawson DBE, KPMG professor emeritus of
management studies, Judge Business School,
University of Cambridge

"In *Stress-Proof*, Dr. Mithu Storoni focuses on the problem and solutions for the reason that 80 percent of my patients decide to see me: stress. Recognizing and managing stress, as described in expert detail in this breakthrough book, can start a revolution in healthcare by focusing at its root causes. Highly recommended."

—Joel Kahn, MD, FACC; clinical professor of medicine and founder, Kahn Center for Cardiac Longevity

STRESS-PROOF

STRESS-PROOF

The Scientific Solution to Protect
Your Brain and Body—and Be
More Resilient Every Day

◎

Mithu Storoni, MD, PhD

A TarcherPerigee Book

tarcherperigee

An imprint of Penguin Random House LLC
375 Hudson Street
New York, New York 10014

Most TarcherPerigee books are available at special quantity discounts for bulk purchase for sales promotions, premiums, fund-raising, and educational needs. Special books or book excerpts also can be created to fit specific needs. For details, write: SpecialMarkets@penguinrandomhouse.com.

Library of Congress Cataloging-in-Publication Data

Names: Storoni, Mithu, author.
Title: Stress-proof : the scientific solution to protect your brain and body—and be more resilient every day / Mithu Storoni, MD, PhD.
Description: New York, New York : TarcherPerigee, [2017] | Includes bibliographical references and index.
Identifiers: LCCN 2017009893 (print) | LCCN 2017022144 (ebook) | ISBN 9781524704087 (ebook) | ISBN 9780143130475 (hardback)
Subjects: LCSH: Stress management. | BISAC: SELF-HELP / Stress Management. | SCIENCE / Life Sciences / Neuroscience.
Classification: LCC RA785 (ebook) | LCC RA785 .S757 2017 (print) | DDC 155.9/042—dc23
LC record available at https://lccn.loc.gov/2017009893

Printed in the United States of America
1 3 5 7 9 10 8 6 4 2

Book design by Katy Riegel

Dedicated to "progress."

*The piston of civilization,
the inaugurator of chronic stress.*

Contents

Introduction:
Unboiling the Egg

THERE IS A peculiar-looking building that stands proud in
the middle of London. It is a *post*-modern building—its
sense of style is so evolved that modernity for it is too passé. Existing skyscrapers already hold enormous legions of humans but
this colossal structure had no intention of stopping there. The
building is wider at the top than it is at the bottom to hold even
more people. Fortunately, rents tend to climb, the higher up you
go. The building is shiny, with glazed, panelized, aluminum
cladding reflecting all who dare shine upon it. Its vulgar vanity
rises high above the level of its peers, so it greedily gorges on
pure, unblemished sunshine while resembling a gargantuan,
Stone Age walkie-talkie.

Londoners were prepared to overlook what many regarded as
an eyesore, until one warm and sunny summer day in 2013 when
for two hours in the afternoon, the building metamorphosed
from a postmodern blunder into a New Age villain. With a light
beam of destruction, it melted cars, blasted bicycles, blistered
paint, and even set fire to a doormat. What really got London

talking, however, was its ability to cook an egg. A journalist placed an egg on a frying pan and put the frying pan at the precise spot on the street below where the south-facing façade of the building converged the sun's rays to temperatures of 243 °F. The egg sizzled and was cooked within moments.

A newly hatched egg, underneath its hard shell, is runny and free. It goes through drastic change when you surround it with heat. There is little to notice on the outside while the inside is altered beyond recognition. We humans are very much like eggs. When plunged into the heat of our lives, our shells may remain the same, but our brains undergo structural change. We call the heat "stress."

The walkie-talkie building in London is a metaphor for modern life. It is a product of globalization, born out of the urge to maximize income, boost productivity, out-compete rivals, leave behind peers, and ambitiously maintain pace with the fast sprint of modernity. In trying to do all of these things, it sends out cross fire. That cross fire cooks the egg. Our brains suffer.

This would be a sorry state of affairs were it not for a recent discovery. It is in fact possible to "unboil" an egg.

The scientist who managed to achieve this earth-shattering feat sent shockwaves through kitchens across the world, left Michelin-starred chefs from New York to Tokyo scratching their heads, and shattered the common saying that it is "impossible to unboil an egg" into extinction. He was quite deservedly awarded an Ig Nobel Prize.

Just as one can "unboil" an egg, one may also be able to "unboil" the brain. This book takes the process of stress and works it backward so, like unboiling an egg, you can "undo" and try to prevent some of the changes that take place in your brain and body when you are placed under the beam of the thoroughly modern and potentially damaging walkie-talkie building that is your daily life.

⊚ LOOKING AT STRESS IN A NEW WAY

I was born into a family of doctors, thinkers, exercisers, and yogis and grew up surrounded by tales of strange and unusual feats: men voluntarily disappearing into the icy Himalayas and living without warmth and with little food, just to train their minds, bodybuilders lying on beds of nails to train themselves not to perceive pain, yogis slowing their heart rate to such a low level that people around them grew frightened they were about to pass away. The brain, I was told, had the power to overturn decisions being made lower in the chain of command. We operate on an autopilot program, called the *autonomic nervous system*. This system is headquartered in the brain but holds great influence over the entire body. It keeps the heart beating and the lungs breathing, even when we forget they exist. In broad terms, half the system is responsible for the stress response and the other half calms us down. The half that triggers stress is known as the *sympathetic nervous system*. I learned how the beguiling effect the mind had on the body held fascination for great athletes. Sir Roger Bannister, who was the first person to run a mile in under four minutes, at Oxford's Iffley Road Track in 1954, devoted his entire career to studying the autonomic nervous system.

I had relegated these marvels from my childhood to one of the less visited bookshelves of my mind until I came across the story of the Dutch explorer Wim Hof, known as the "Iceman." In 2007, Hof climbed part of Mount Everest wearing shorts and shoes; he completed a marathon at temperatures of around −4 °F, similarly attired, two years later. On January 26, 2007, Hof secured the world record for running a half-marathon while barefoot on ice or snow, at 2 hours, 16 minutes, and 34 seconds.[1]

More recently, Wim Hof's help was sought for an experiment to test an intriguing concept.[2] Is it possible to override the body's response mechanisms to a bacterial invader by training the mind? In other words, was it possible to voluntarily control the autonomic nervous system? Can you turn up the sympathetic nervous system volume "on command"?

Wim Hof put twelve healthy volunteers through a ten-day training regimen involving meditation, breathing exercises, yoga, and cold exposure in an attempt to acquire the ability to voluntarily activate the sympathetic nervous system. After the training period, both the volunteers and a control group were injected with bacterial endotoxin at levels likely to spark an immune response and cause illness. Half an hour before the injection, the trained volunteers were "commanded" to voluntarily activate their sympathetic nervous system (a feat traditionally deemed impossible). They did. As a result, the trained volunteers had higher levels of epinephrine (released during the stress response) circulating in their bodies when the endotoxin entered their blood. The epinephrine made their bodies produce more of the protein *IL-10* in response to the endotoxin, compared to the control group. IL-10 has an anti-inflammatory effect and the trained volunteers had fewer flu-like symptoms and recovered faster both from their fever and from their stress response to the endotoxin. The study, though small and the first of its kind, spectacularly demonstrated that contrary to traditional belief, it is entirely possible to exert some degree of voluntary control over the autonomic nervous system—and hence even the immune system—bridging the gaping abyss hitherto believed to separate the body from the mind.

As an undergraduate I was intrigued by the malleability of the brain when I learned how Nobel Laureates David Hubel and

Torsten Wiesel had demonstrated that the brain begins as a tabula rasa block of blank marble and the world sculpts itself upon it. If kittens are never shown horizontal lines by the world, their brains can't recognize them when they are grown-up cats. I also observed the malleability of the human spirit on the wards as a medical intern and then as a medical resident. There are patients with odds stacked heavily against them who pull through while others with the odds in their favor do not. There are those whose illnesses chronicle their mental state so well, the two can be plotted as a perfectly straight line on a graph, over the course of months and even years. The private thoughts of men and women at the sunset of their lives seem to decide whether they will live or die at the next fork in the road. And, of course, there is the infamous placebo effect, which has been known to accomplish near-miracles.

During my medical internship, I developed a mild autoimmune condition that I was desperate to get rid of. It served as an annoying antenna for the level of stress in my life. The moment the stress intensified, so did the condition. I lived with my antenna until I took up hot yoga as a hobby while I was studying pupillometry in London. Pupillometry is a niche specialty within the field of neuro-ophthalmology, dedicated to the study of pupil movements. How fast the pupil gets bigger, what it looks like, how quickly it shrinks, and its delicate microscopic flutterings are endlessly fascinating if you appreciate that the pupil is a direct window to the autonomic nervous system. The pupil dilates when the sympathetic input reaching it is intensified, which is why your pupils look big if you are stressed. I discovered over the course of a few months of practicing hot yoga that my own baseline pupil measurements appeared to change, suggesting the possibility of a reduction in my baseline sympathetic nervous system activity. In parallel with this observation, my autoimmune condition seemed to shrink, too, until it disappeared completely.

It became apparent to me that it wasn't so much the power of the mind, as the power of *looking after the mind* that seemed to be the protagonist. I wasn't *thinking* myself better; I was making my mind's baseline state healthier by training it, feeding it, nurturing it, and resting it. When the mind is in its optimal state, it reacts differently. It is more resilient in the face of stress. It heals faster after a trauma. It thinks constructive thoughts and views the world rationally. It raises the threshold for pain perception, bolsters the immune system, and slows the process of any disease. It truly makes us more *stress-proof.*

When the mind is in its optimal state, it reacts differently.

℗ TOWARD STRESS RESILIENCE

János Hugo Bruno Selye was a legendary Austro-Hungarian physician who is recognized today as the father of stress research.[3] In 1956, Selye described stress as "a scientific concept which has received the mixed blessing of being too well known and too little understood."[4] We have come some way since then, though we have an even longer way to go.

You operate on a set point. That set point is maintained by your intelligent brain primarily through your autonomic nervous system and its sympathetic and parasympathetic activity. If you enter a hot room, you sweat. If you drink too much, you go to the bathroom. If you lie down flat and the pressure in your brain rises, your body may reduce your blood pressure. Your

body is programmed to meet change with constancy. If your world changes, your body sets off mechanisms that work to keep you at your set point.

Stress is when your brain and body *change in response to change*.[5] Your changing world, to which you are unadapted, makes you change your set point.

Your blood pressure may be ideal for lounging around, but if you expect to be attacked by a lion any second, it could do better. If that lion assaults you and your wounds bleed profusely, your blood will not have enough pressure to reach the brain, nor to supply your muscles to help you run away. If your blood pressure were set at a higher baseline, the drop in blood pressure from severe blood loss would not be quite as debilitating and you would be able to stay alive. Normally, your body tries to keep you at one "set point" baseline blood pressure. In stress, your body changes your set point and sets it to a higher level to anticipate and prepare for a possible drop in blood pressure.

Your brain changes the set points of a range of variables, to raise the odds of success in the face of imminent danger. Once that danger has passed, your set points are reset. If the threat *never* passes, or is *too frequent*, your set points fail to change back. That's when you suffer from the negative effects of chronic stress. Your blood pressure *stays* raised. Your stress signal *stays* turned on. Your stress signal is mediated by your sympathetic nerve network, and sympathetic activity stays elevated. The stress signal generates stress hormones and these may persist.

The reason changing set points leads to illness is that we have evolved to operate at our normal set points. We are able to tolerate altered set points—elevated stress levels—only for short bursts of time. Staying at those set points can damage the brain and the body. If you are attacked, you won't care much for the short exposure to high blood pressure or to high blood sugar–related

damage, because the good they bring in saving your life far out-
weighs the comparatively smaller harm they may cause. If you
are not being attacked and your blood pressure and blood sugar
level stay high, both may cause damage, without doing any good
at all.

A persistently high stress signal can show up as a chronically
high blood pressure.[6] Globally, the incidence of high blood pressure is rising, prompting some to wonder if this is a direct result
of a global increase in stress caused by urbanization and globalization. It is possible that the global rise in insulin resistance and
type 2 diabetes also stem from the same root.

As set points change, the wiring within the brain also changes.
The brain is programmed to optimally function in a non-stressed
world that is peppered with occasional bouts of stress. If it is made
to exist in a stressed world that is peppered with occasional bouts
of non-stress, it tries to change its connectivity in order to function optimally in this novel paradigm. The change brought on
by chronic stress is an adaptation response, or rather a *mal*-
adaptation response, because the change
does not lead to *better* adaptation. Chronic
stress dwarfs our lives rather than enabling
us to thrive in our environment.

*Chronic stress
dwarfs our lives.*

☺ A MULTIFACETED APPROACH

Stress remains as much a conundrum today as it was a hundred
years ago, with the difference that where before we saw a tangled
mass of strings, today we are able to identify many of the strings
that make up the tangle. Each area of malfunction is one string.
If you do all you can to look after each string, they are less likely
to tangle into an unmanageable mass.

In the landscape of brain health, a multifaceted approach—like the one presented in this book—is emerging as a better alternative to one that only resolves a small piece of a complex conundrum.

In the chapters that follow, I describe seven areas of malfunction likely to be experienced by anyone who is chronically stressed: weak control over attention, too much or too little cortisol, altered synaptic plasticity, an out-of-tune body clock, inflammation, insulin resistance, and flagging motivation. One person may not be affected by *all* these things but *most* people will be likely to suffer from at least one of them. I take the approach that if you target each of these areas to minimize your chances of succumbing to each malfunction, you may overpower the effect of chronic stress. If you stringently maintain healthy brain activity, a regularly ticking body clock, and quash the slightest smoke of inflammation, if you train your attentional skills and learn how to regulate your cortisol levels, if you keep your motivation on track and do all you can to diminish your risk of insulin resistance, you will put up a strong fight against chronic stress. This book tells you how to navigate through each of these channels. I have tried to cite the scientific evidence behind each suggestion, and where possible, I have cited results from randomized, controlled trials. I've also used the latest findings from relatively nascent research fields, where results are preliminary but hold promise.

The interventions described in this book are likely to bring benefits to everyone. You won't wake up wearing a red cape overnight, but over time you will turn yourself into the *best version of you.*

STRESS-PROOF

CHAPTER 1

The Two Sides
of Your Brain

THE MOMENT YOU encounter a situation that has the potential to be stressful, two conversations take place inside your
head. Your entire world can look very different, depending on
which one you choose to attend to. One conversation is rational,
calm, and reasoned. The other is emotional, impulsive, and
hasty. Chronic stress raises the volume of the second and mutes
the first.

A PREFRONTAL EXECUTIVE

Your intelligent brain is like a giant corporation with a multi-
tude of departments and subdepartments. An unimaginable
amount of information passes through it and must be appropri-
ately processed. The corporation's aim is to adapt your behavior
to reap maximal benefits from the environment you are in. At
the head of your corporation sits the chief executive officer, who

coordinates the corporation's machinery and decides which departments should be accentuated and which should be toned down. This process must be able to adapt to changing circumstances. If the CEO orchestrates the corporation with wisdom and precision, it will likely thrive.

At the front of your brain, behind your forehead, sits a region known as the **prefrontal cortex**. It plays a central role in executive control and behaves, with collaborative assistance from others, like a CEO.

In every situation, it carefully assesses your terrain and formulates the best possible strategy for navigating through it. It modulates and controls activity across your brain's various departments to create as favorable a climate as possible for you to accomplish what you are doing. For example, if you are trying to read a long e-mail in the middle of a hectic, noisy office, it coordinates networks so the noise is muted and the distractions emanating from your surroundings are dimmed, so that you may focus.

Like any great executive, it has some special talents. It gathers as much information as it can from your present situation and holds on to it as *working memory*. Your working memory lets your prefrontal cortex relate what is happening now with what just happened a moment ago, so it can predict what will happen next and modify its strategy for you, if needed. Your prefrontal cortex also controls the spotlight of your attention and decides where that spotlight should be beamed. It scrutinizes data coming in from multiple channels to decide what deserves attention and what does not. The scrutiny takes place at many levels within the prefrontal cortex and includes analyses of analyses as well as consultation with long-term memory stores via a region of your brain known as the *hippocampus*. If your thoughts and sensations are unimportant and irrelevant to your task at hand, your prefrontal cortex lowers their volume so your attention does not waver from what you are doing.

Your prefrontal cortex, in association with other networks, strategically plans, reasons, regulates behavior, makes decisions, and exerts top-down control over other parts of your brain as you navigate toward a goal. Through trial, error, and intelligence it learns to assign an appropriate behavior to a given set of circumstances and to improve upon that behavior as soon as new information presents itself. It is *always* learning and trying to upgrade its intelligence.

This puts your prefrontal cortex in a state of intense activity as new networks form and change and new connections between brain cells (synapses) materialize and weaken. There are an unimaginable number of synapses within your brain, and these synapses shift and change all the time in a state of heavy flux. The evolving change in the strength and activity of synapses is known as *synaptic plasticity*. When we adapt to a new situation and the brain rewires itself to cope, it relies heavily on synaptic plasticity.

ⓔ THE RATIONAL REGULATION OF EMOTION

Your emotional instinct is a valuable tool for navigating the nuances of today's urban environment because your threats tend to come from social interactions, rather than from wild animals. When these threats set off negative emotions in your mind, they may bypass careful analysis by your prefrontal cortex, to save you time. This short circuit prevents you from filtering out false alarms.

The brain circuitry that processes your emotions is extensive and includes both positive and negative feedback loops. Many of these loops are modulated by parts of your prefrontal cortex. If your prefrontal cortex decides it is in your best interest to stay on high alert for possible threats coming from your environment, it may raise the volume of your emotional response. If it decides your emotions are proving distracting to what you are doing, it may mute your emotional response and shift the spotlight of your attention away to a worthier target. If it malfunctions, your emotional response may be disproportionate to what your situation warrants.

One of the main players in your emotional network is the *amygdala*. The amygdala carries out a quick preliminary scan of the information coming in from your environment and then sends signals to various other parts of your brain including the prefrontal cortex. The prefrontal cortex, in turn, sends signals to your amygdala. Parts of it may promote or demote its activity depending on the other information it has gathered. During emotional conflict, for instance, parts of the prefrontal cortex appear to "restrain" the amygdala.[1] One other key team member in your emotional network is the (mostly ventral) *hippocampus*, which collaborates with your amygdala.

The prefrontal cortex plays an essential role in conducting the orchestra of your brain so your response to the world you find yourself in is always rational and reasoned. If the regulatory skills of your prefrontal cortex are hampered, your response to your environment may be irrational and inappropriate and your experience of your life will change. Brain scan images of people suffering from chronic occupational stress or post-traumatic stress disorder (PTSD) show signs of defective prefrontal regulation of emotion and behavior. Not being able to down-regulate negative emotions is associated with burnout.[2] [3]

If your emotions are not regulated, negative emotions may

surface easily and take over your mind. You may view the world with a negative bias, dwelling on its negative features and remembering negative experiences more than positive ones. Your unregulated perspective may make the world feel uncertain and unpredictable, so you constantly feel on edge and anxious. Each of these intense negative emotions feeds back into your brain's emotion networks, amplifying and propagating their activity.

Here is an example of how the presence or absence of good emotional regulation can give you a drastically different experience of the same event.

Life through an Emotional Lens:

Your usually cheerful boss did not smile back at you this morning. You arrive at your desk and start wondering why. You have poor control over your emotions, so your mind creates worst case scenarios as you interpret everything you see and hear with a strong negative bias. You feel anxious and guilty. You worry that your boss has some bad news to share with you, in light of rumors of cuts in your company's budget. You notice others around you discreetly smiling to themselves and wonder if they are laughing at you because your sacking is an open story. You have bills to pay and that new mortgage. You panic.

Life through a Rational Lens:

Your usually cheerful boss did not smile back at you this morning. As you rationally try to understand why, your prefrontal cortex and hippocampus carefully revisit the scene and go through their inventory of past experiences. They come upon a distant memory buried in your mind, of someone gossiping about your boss's sudden interest in Botox therapy. Thinking back, you now remember how the edges of her eyes twitched while her mouth and forehead remained frozen. Looking around the office floor, you notice your colleagues trying (and failing) to keep a straight face. You smile, too, at the hilarity of the situation.

⊚ YOUR AUTOPILOT RESPONSE SYSTEM

Your autopilot nerve network, the autonomic nervous system, rapidly carries signals from your brain to your body. Its two halves, the *sympathetic* and the *parasympathetic*, which work together to keep your body's engine running smoothly, stay permanently switched on. When their activities need adjusting, their gain or *tone* is changed without turning them off completely. For instance, sympathetic input to your heart *raises* your heart rate whereas parasympathetic input *lowers* it. If your heart needs to beat faster, the tone of the sympathetic input is raised and the tone of the parasympathetic input is reduced, but both continue to provide input to the heart.

When your body experiences stress, your brain sets off two distinct chain reactions. The first chain involves your fast-acting autonomic nervous system whose two halves work in opposition during stress. Your sympathetic tone rapidly rises and your parasympathetic tone falls and this chain culminates in the release of epinephrine (also known as adrenaline) and in a range of physiological responses such as rapid breathing, a quickened pulse, and heightened alertness. The second chain of events begins in your hypothalamus and ends with the adrenal gland releasing the stress hormone cortisol. The two chains interact and feed forward and back upon each other until the stressful moment is passed. At this point, your parasympathetic tone rises and your sympathetic tone falls. Your parasympathetic network becomes more active as you relax and your sympathetic network becomes more active when you are stressed.

Your amygdala and its close collaborators (known as the Central Autonomic Network) are intricately connected to the circuitry of your stress response. This explains why things that affect you emotionally can rapidly trigger a stress response.[4] Experiences that elicit intense negative emotions can increase your sympathetic tone whether they come from the world around you or from the thoughts floating through your mind.[5] If you experience many such emotional triggers, or if you cannot regain control over your emotions quickly after they have been aroused, you will be prone to frequent bouts of stress and your sympathetic tone may stay unnecessarily raised.

Your brain ignites a stress response when it thinks you are being threatened. That threat may be physical or emotional. The stress that we tend to experience most of the time in today's industrialized, urbanized world takes the form of psychosocial stress which acts through your emotional reactivity. Since your prefrontal cortex regulates your emotional reactivity, it plays a vital role in your susceptibility to stress.

Its critical role becomes apparent when you find yourself in an unexpected, stressful situation when it reins in your stress reactivity, regulates your emotions, and keeps your attention fixed on the task at hand. If it does its job well, it softens the stressful impact of the situation. In the moments that immediately follow a stressful experience, your prefrontal cortex shifts your attention away from the inflammatory thoughts simmering in your mind, so you can recover as quickly as possible and move on. If it is not able to regulate your emotions, your recovery is slow and may even be incomplete.

◎ A WORD ON TERMINOLOGY

The network of brain cells that process emotions is vast and complex and I will be referring to it as the *emotional brain*, so as not to overburden you with too many technical terms. Similarly, the *rational brain* refers to the networks in the prefrontal cortex that are involved in goal-directed decision-making and behavior, emotional regulation, working memory and learning, and attentional control, and are generally responsible for making sure you respond to your environment as rationally and wisely as possible. This term also includes networks in the (mostly dorsal) hippocampus, a key player in learning and memory, as well as in some other regions that may collaborate with the prefrontal cortex. The terms *emotional brain* and *rational brain* in the context of this book refer to what the networks *do in the context of psychosocial stress*. In reality, the brain is not anatomically divided into rational and emotional sections, and emotion and cognition are tightly interwoven and served by overlapping circuits.

◎ ACUTE AND CHRONIC STRESS

If you imagine the brain as an orchestra conducted by the prefrontal cortex, the melody being played is synchronous and balanced *most* of the time. During acute, uncontrollable stress, the conductor, your prefrontal cortex, gives the floor to one instrumental section that emerges out of the harmonious symphony and powerfully dominates the stage. This section processes your negative emotions. After the stressful experience is over, the prefrontal cortex shifts attention away from this section and the soothing harmony resumes.

The brain of an adult human alters in response to what is *asked of it*. It rapidly adapts to meet changing demands so it can thrive in a dynamic environment. If its experience of stress is no longer acute but *chronic*, it may change its connectivity and structure to adapt to this new setting. The temporary weakening of prefrontal control over emotions and behavior now persists and dysregulated emotional behavior lingers on. The change in connectivity makes the imbalance between rationality and emotional reactivity long-standing instead of temporary.[6] Many of the manifestations of chronic stress, from impaired emotional regulation to changes in motivation, behavior, and the ability to feel pleasure, may be the result of diminished prefrontal control.

There are many routes by which chronic stress can progressively weaken prefrontal control networks. The prefrontal cortex and the hippocampus are both in a state of constant flux and they rely heavily on intense synaptic plasticity. Any process that obstructs this activity, such as chronic stress, interferes with their functioning. Interestingly, the prefrontal cortex and hippocampus are also vulnerable to damage from aging and degenerative diseases such as dementia.

Brain cells in the prefrontal cortex (known as pyramidal cells) have a shape that resembles a tree. They have branches that extend outward (known as dendritic branches). These branches are involved in synapse formation. Chronic stress makes these branches regress. It also affects cell-to-cell communication and hinders coordinated electrical oscillations between brain cells, which are vital for information processing.[7] These effects compromise the ability of the prefrontal cortex and the hippocampus to do their job properly, and regulatory control may suffer.[8] [9] [10] [11]

The poorly regulated emotional brain may now react more easily. While dendritic branches may regress in the prefrontal cortex, they may *grow* in the amygdala.[12] A recent study has shown an inverse relationship between feeling chronically stressed and the size of the prefrontal cortex.[13] Each small bout of acute stress a chronically stressed brain encounters may feel more intense and prolonged than it otherwise would.

We become what we behold. We shape our tools, and thereafter our tools shape us.   —*Marshall McLuhan*

A situation that would not have ruffled your feathers before now raises your blood pressure. As it suffers damage from chronic stress, the prefrontal cortex starts performing poorly in tests of working memory and cognitive flexibility. It loses its grasp over attention and self-control. Your experience of the world becomes less balanced and you may find yourself dwelling on negativity when your mind wanders, and jumping to negative conclusions in moments of doubt.[14] It might feel increasingly difficult to disengage from negative emotions and negative thoughts to focus on your task at hand. Eventually, this negative spiral of chronic stress may culminate in depression.

A high sympathetic tone is thought to play a part in hypertension, obesity, and insulin resistance, which are all globally on the rise.[15] [16] A skewed sympathetic/parasympathetic balance can also significantly affect your heart's ability to adapt to a new situation, increasing the risk of a heart attack. Preserving the health of your rational brain can potentially extend your life.

✪ MOVING FROM ACUTE TO CHRONIC STRESS

If your rational brain were to stop conducting the orchestra of your brain's networks, their beautiful harmony would turn into a chaotic cacophony. Behavioral and emotional regulation would collapse. Your emotional brain would be inappropriately reactive. Your experience of pleasure and pain, reward and failure, would be grossly distorted. Chronic stress cripples your rational brain's ability to skillfully conduct its orchestra. As the damage from chronic stress accumulates and your brain's connectivity changes, you react vigorously to trivial situations that ought not to be stressful. In this way, chronic stress increases the number of stressful episodes in your day-to-day life.

Chronic stress also affects your recovery from stressful episodes. Your brain may take longer to recover after each episode and there may not be enough of a gap before the next episode is encountered. In this way, chronic stress potentiates itself. While stress inflicts its damage in this way, unwise lifestyle choices, diet, sleep patterns, and behavior steadily contribute to the damage load and reinforce the negative effects of chronic stress on your brain.

Your rational brain's robustness sits in the eye of the storm of chronic stress. It lies at the core of our stress strategy.

◎ CREATING A STRESS STRATEGY

When your brain triggers an acute stress reaction, it sets off one chain reaction after another until these eventually affect almost every system in the brain and body to bring on a change. For the purposes of the stress strategy in this book, I have chosen to focus on seven of these changes that take place during acute stress:

1. The emotional brain is on high alert.
2. You release an appropriate amount of stress hormones.

3. There may be an increase in synaptic plasticity and *in mice, the birth of new brain cells.*
4. Body clocks temporarily malfunction.
5. You become inflamed.
6. You are temporarily resistant to insulin.
7. You suddenly feel motivated.

These seven operations are like seven "stress agents" who form a shield to protect you from imminent danger. Your alerted emotional brain keeps you vigilant for clandestine attackers. Anger motivates you to stand your ground. Cortisol and other hormones and messengers help you to defend yourself. Inflammation protects you from germs that deviously enter your body through your wounds. Your body clock's openness to adjustment makes you adaptable. The insulin resistance prevents your body from guzzling down the sugar in your blood, so there is plenty left for your brain. Your motivation keeps you going and exterminates hesitancy. Enhanced synaptic plasticity helps you to learn from this experience. These seven mechanisms cease as soon as the stressful moment has passed.

A **chronically stressed** body also shows signs of these seven changes, but a closer look reveals that though the changes *seem* identical, they are different. It is as if each of the above agents has done its job and then "turned rogue":

1. The emotional brain may *remain* poorly regulated.
2. You may *release too much or too little* cortisol in response to stress.
3. There may be a *decline* in synaptic plasticity. *In mice, new brain cells cease to be born.*
4. Your body clock may *permanently* malfunction.

5. You may *remain* inflamed.
6. You may be *chronically* resistant to insulin.
7. You may have a *persistently* altered sense of motivation—pleasure and reward.

If you study each route closely, you will notice that aspects of diet, lifestyle, and behavior can push you toward illness along each route, *independently* of stress. For instance, one of the routes is chronic inflammation. Unwise dietary choices can incite chronic inflammation whether you are stressed or not. Another of the routes follows the body's clock. If you have erratic sleeping patterns and do shift work, your circadian rhythm will suffer even in the absence of identifiable stress. If you do all you can to live in a way that minimizes inflammation and that keeps your circadian rhythm finely tuned, then you will be able to resist and even overturn the efforts of stress to lead you toward illness along those routes.

Our strategy for stress resilience will focus on addressing each one of the seven routes. You are building a solid defense by going on the offensive along each route and identifying and eradicating aspects of your diet, lifestyle, and behavior that drive you toward illness. When stress tries to push you along a route, you will be able to push back with force—and prevail.

◎ SEVEN ROUTES TO BECOMING STRESS-PROOF

As we build our stress strategy through each of the seven routes, our priority is to encourage a return to healthy activity within rational brain circuitry.

Route One: Improving Emotional Regulation

If you are frequently or easily stressed, lose your temper more than you used to, or feel unduly worried and anxious, prefrontal control over your emotional reactivity may be faulty.[17] *Emerging evidence suggests you may be able to improve prefrontal control with specific training techniques.*

Route Two: Getting Stress Hormones in Check

Chronic stress makes you release *too much or too little* of the stress hormone cortisol. Abnormal levels of stress hormones cause harm to the structure and normal function of the brain cells and glial cells that form the substance of the rational brain's networks.[18] *With specific modifications to lifestyle and behavior, you may be able to even out irregularities in stress hormones and messengers circulating in your blood during and after a stress response.*

Route Three: Encouraging Healthy Activity in the Rational Brain

As the rational brain records memories, learns, and creates strategies, new networks form with novel synaptic connections while existing networks are modified. Chronic stress stunts these processes.[19] *If you take steps to enhance these processes by actively stimulating growth and activity in your rational brain, you may be able to counter effects of chronic stress.*

Route Four: Tuning Up Your Body Clock

Every department and sub-department inside you works in time to a clock. There are thousands of clocks distributed across your brain and body, which are regularly tuned by entrainment signals. These signals are affected by your habits and lifestyle. If your clocks malfunction, biochemical and physiological processes across your entire body go off-course. Chronic stress disrupts your clocks and your brain is especially vulnerable to damage from disordered biorhythms.[20] *Keeping your clocks precisely tuned with specific dietary, lifestyle, and behavioral modifications will help offset the chaos inflicted by stress.*

Route Five: Taming Chronic Inflammation

Inflammation can interfere with synaptic plasticity and lead to inappropriate and aberrant synaptic connectivity.[21] *Specific dietary and lifestyle modifications can minimize baseline inflammation, reduce intestinal permeability (a contributor to inflammation), and provide a buffer against stress-induced inflammation.*

Route Six: Fighting Insulin Resistance

Insulin affects neurotransmission and synaptic plasticity. Synaptic plasticity and network activity in the prefrontal cortex and hippocampus rely heavily on energy supply, and they suffer if glucose levels are poorly regulated.[22] *You can reduce the risk of developing insulin resistance with strategic changes to your diet, exercise regimen, and lifestyle.*

Route Seven: Boost Motivation—Pleasure and Reward

When your prefrontal cortex regulates your behavior, it modulates the circuits that process your sensations of pleasure, reward, and motivation. If this process malfunctions, your experience of pleasure can be adversely affected. Anhedonia can be brought on by chronic stress.[23] *Modifications to your behavior and lifestyle can help to protect your reward circuitry, and keep you feeling driven, upbeat, and positive.*

Imagine your stress-proof barrier as a brick wall. Every brick in that wall is important. One brick may look after the health of

your gut, while another might tune the clocks that sit in your
liver. Each brick brings benefits on an individual level, but it is
when you coordinate all the bricks and synchronize their individual powers to create a solid wall that
the effect becomes formidable. Protecting

Protecting your- yourself against stress requires many small
self against stress changes. The small changes have a syner-
requires many gistic effect when implemented together.
small changes. Harnessing their synergy creates a powerful fortification against stress.

Let's begin!

CHAPTER 2

Bolstering Emotional Regulation

⊚ WHAT IS THE *POINT* OF STRESS?

Imagine you are given a dose of an anesthetic that prevents you from feeling pain. You are fast asleep on your sofa at home. Suddenly, an intruder with a machete strikes your right leg and runs off. Blood spurts from the wound. Although you are unable to feel pain on account of the anesthetic, you have sensors in your body that monitor your blood flow. They are activated *after* you have lost a significant amount of blood. They frantically try to correct the situation but it is too late. They fail. You die. What could have saved you? *Time.*

If you had not had the anesthetic and had felt *physical pain* the moment the machete struck your skin and *before* your wound started bleeding, the pain would have set off a *stress response.* Your body's emergency system would have been activated sooner. This would allow just enough blood to supply your brain so you could crawl to your cell phone and call 911. This rescue mission is *one step* ahead of waiting to lose blood.

If you *sensed* an intruder somewhere in your home even be-fore you saw him, your emotional brain would react violently and set off a *stress response*. It would immediately start preparing you for the eventuality that the intruder will attack you with his ma-chete. You would race to find your cell phone and call the police and would either arm yourself with a weapon or find a good place to hide until the police arrived. This rescue mission is *two steps* ahead of waiting to lose blood. It is *one step* ahead of waiting to feel pain. Although both pain and loss of blood set off the stress response, your emotional instinct gives you the power to *predict* pain before it occurs. It buys you time.

Your emotional brain has evolved to save your life by speed-ing up your rescue mission by three notches. Instead of reacting *after* you have been attacked, it helps you *predict* the attack and *prepare* for it as best you can.

Of the many routes that lead to a stress reaction, the most com-mon encountered in today's urban age is psychosocial stress—everyday encounters with rude strangers, demanding supervisors, and unexpected traffic on your way to work. The point of entry for these stressors is through your emotional reactivity. In the past, this would have bought you time and saved your life. Today, it en-dangers it. A volatile emotional brain that sees a small ripple as a tsunami and feels an elbow nudge as an earthquake will turn a rainy day into a tempestuous one. Over time, an overactive emo-tional brain has trouble bouncing back. It's much easier to get back to normal after a week of rain than after a week of cyclones.

© EMOTIONAL REGULATION

Your emotional brain is your psychosocial stress "tinderbox." If it catches fire, it starts a stress reaction. Chronic stress lowers your emotional brain's threshold for sparking a stress response. You need to prevent that tinderbox from catching fire.

The moment something provokes your emotional brain, you might suppress it, override it, ignore it, or distract your attention. It takes immense self-control to be able to do these effectively.

Becoming better at self-control and self-regulation can help you to control negative emotions when they are ignited. Those who have good self-control tend to cope better with stress. They perceive fewer situations as being stressful and they react with less intensity when faced with a stressor.[1] Brain scan studies have identified at least one region within the rational brain that plays a role in self-control, the **dorsolateral prefrontal cortex** or **dlPFC**.[2] It is also involved in **self-regulation**, which is a broader form of self-control. Long-term work stress is associated with a shrunken dlPFC.[3] Damage to the dlPFC can make people more vulnerable to depression.[4]

People who meditate find it easier than non-meditators to bestow their complete attention upon an object. Improving your ability to shift your attention away from something that is causing you emotional distress and redirect it toward something pleasant or distracting can help with emotional regulation. Prolonged concentration and focus demand a degree of self-control.

Another strategy that might help to prevent setting the emotional tinderbox alight is to force yourself to look at a provocative situation in a new light that makes it feel less distressing. This is known as *cognitive reappraisal*. A robust rational brain may help with cognitive reappraisal.[5]

A healthy brain will usually use a combination of strategies to regulate emotion, and often the magnitude of what you are facing will dictate the strategy you might use to suppress your emotional brain. If someone has pushed against you on the street, you might try to reinterpret their action as not having been de-, liberate and this will pacify your anger. If you have witnessed a horrific accident, your mind will try to block out what you have just seen, because it is impossible to reinterpret it in any other way.[6]

Our stress strategy for emotional regulation includes both short-term and long-term fixes. The short-term angle will focus on strategies you can use to restrain negative emotions as they surface. The long-term angle will focus on ways to train yourself to improve your overall ability to regulate negative emotions over time.

◎ SHORT-TERM FIXES

Rationality and emotional reactivity are almost mutually exclusive under stressful conditions. This is why you find it difficult to concentrate and logically think through problems when you're upset. Keeping your rational brain tightly engaged will make it more difficult for it to release its attentional control and let your emotions take over.

Playing Games

If something has just upset you and you want to stifle an onslaught of negative emotions, you need to beam your spotlight of attention away from your emotions. Engaging your rational brain as completely and as quickly as possible with a task that needs its unwavering attention can help. One way of doing this is by playing games on your smartphone that test your working memory or reasoning skills and absorb your attention in the process.

In one small randomized, controlled study, eleven healthy volunteers were made to feel sad by recalling negative memories and listening to somber music. They were then made to perform a working memory task or to play the game *Tetris*, a spatial reasoning game, while their negative mood was perpetuated. The working memory task required the volunteers to add a number appearing on a screen to a number that had been presented before. A control group spent an equivalent amount of time looking at a cross on a computer screen. A brain scan revealed that the working memory task and *Tetris* both suppressed the emotional brain's reactivity. There was reduced activity in the amygdala in both settings, compared to the control condition of looking at the cross on the screen.[7]

- *If you feel distressed after a stressful encounter, immerse yourself in a spatial puzzle like* Tetris *or a working-memory game. Play until you lose yourself in the game and are able to transiently forget what just happened.*

Immerse Yourself in Flow

The concept of *flow* was first coined by the great Hungarian psychologist Mihály Csíkszentmihályi. Flow denotes a state of being so completely absorbed in an activity that any unrelated thought or sensation is obliterated. The activity must be challenging enough for you to have to engage your rational brain, but not so challenging that you lose interest because it feels too difficult or is too stressful. Ideally the activity should involve a continuous series of small challenges, each of which can be overcome with some effort. Overcoming one motivates you to try to overcome the next, and the effort consumes your rational brain.

This continuous series of challenge, success, and motivation garners momentum and your rational brain remains absorbed for a long time. While it is tightly engaged, it is difficult for your emotional brain to surge with unnecessary negative emotions. Negative emotions often surface during boredom; in his book *Flow*, Csíkszentmihályi provides many examples of people inducing a flow state in their work and eliminating boredom. For instance, a person assembling parts in a factory might avoid feeling bored and negative by treating the experience as a challenge to devise a strategy for assembling the parts in record time. Csíkszentmihályi cites yoga practice, music, and sports as examples of activities that can put people into a state of flow.

Entering into a state of flow immediately after a stressful experience may help you avert unwelcome negative thoughts. The longer you spend in a state of flow, the longer you keep the lid on your emotional brain. People who incorporate flow in their daily lives report being happier and less stressed. Flow does not arise from relaxation and rest—it can only be experienced during an activity. If you view your work as an *opportunity for flow* rather than an opportunity for stress, you are likely to feel happier.[8]

If you view your work as an opportunity for flow rather than an opportunity for stress, you are likely to feel happier.

- *Try to enter into a state of flow immediately after a stressful experience.*
- *Create opportunities for a flow state as often as possible.*
- *When you approach a project, optimize it for achieving a state of flow by choosing a challenging way of tackling it even if an easy (and boring) option exists.*

@ LONG-TERM FIXES

Since attentional control, self-control, self-regulation, and cognitive reappraisal all contribute to regulating emotion, becoming better at each of these through practice and possibly also through transfer effects (where improving a skill in one area improves a skill in another) may help you rein in unwanted negative emotions in the long term.

Attention Training

Every time you actively disengage from a thought and shift your attention onto a defined target, you are using the rational brain networks that control where the spotlight of your attention falls. There is some evidence that networks may become "stronger" and more efficient with use. If you use these attentional control networks often, you might be able to *train* them into becoming stronger.

Focused Attention Meditation

One way of practicing repeatedly controlling your attention is through *focused attention meditation*. Focused attention meditation is what happens when you focus your full attention on a defined target. It can be an object in front of you or a picture in your mind's eye. You must block distracting thoughts, and when you are distracted you must rein your focus back onto the object.

Practice at work!

Although it may work better with your eyes closed as you visualize an image in your mind's eye, you can practice focused attention meditation in the middle of your office with your eyes open. Choose an object to focus on. For example, take a mug with a picture on it and place it on your desk in front of you. Set a timer on your phone for two minutes and focus your entire attention on the picture. Whenever your attention drifts away, rein it back in. Focus. Give it all you've got. If two minutes is too difficult, start at one minute. If two minutes is too easy, try five minutes.

Most people practicing focused attention meditation go through an incessant loop of four stages:

1. They focus on the target.
2. Their minds drift off—but they are unaware of this.
3. They suddenly become aware.
4. They disengage from distracting thoughts and shift their focus back to their target.

If you use a brain scanner to study the brains of people while they meditate in this way, you will find the dlPFC, in concert with other cortical regions, lights up during stages 1 and 4, both while they are *reining in* their attention and while their attention stays fixed on the target. With regular practice doing this, the circuits that are involved in focusing attention may become reinforced.[9] The implications of this finding are tremendous. It seems possible to "rewire" your brain with your own thoughts to create a new, stronger circuit that keeps your emotional reactivity under control. Meditators are able to control their thoughts and emotions with greater ease than non-meditators.

Playing computer games that require focus may also improve attentional control. The game *Tetris* is one example.[10]

- *Practice focused attention meditation every day. Use a coffee cup at work.*
- *Play games that require concentration, such as chess and memory games.*

Redirect Attention to the Positive

If you get into a habit of recognizing a negative thought, detaching from it, and actively switching your attention to something positive or neutral, your world will change. Think about how many times you carelessly surrender to a negative thought during an ordinary day. You climb out of bed and draw the curtains to discover that it is raining. Without thinking twice, you use the rain as a conditioned cue to lapse into negativity even if there is no justification for it, because "everyone complains about the rain." You head into the shower and find the towel on the floor, carelessly placed there by your daughter. You dwell for a moment on your daughter's carelessness and feel annoyed. You sit down for your usual coffee and realize you have run out of coffee and must make do with tea instead. You briefly focus on your partner's tendency to forget things from the shopping list and feel frustrated. If a news reporter were to burst through your door to ask you how you feel, your answer would be "miserable."

At the precise moment when you chose each of the negative thoughts, you were at a fork in the road and you could have made a *different* choice.

When you surrender to a negative thought, that thought is likely to introduce another, and you may be drawn into a negativity vortex. Cognitive control requires effort. It is more laborious to detach from an emotional hurricane than from the first negative thought that sparked it. Once your negative emotions overwhelm your mind, it becomes difficult to evict them. Your rational brain needs less strength to *prevent* entry than it does to carry out an *eviction*.

But we can choose to focus on the bright side. When you drew the curtains and noticed the weather, you could have *chosen* to switch your focus onto how lush your parched lawn will look after the rain and how the rain is a welcome respite from the recent warm temperatures. When you noticed the towel on the bathroom floor you could have *chosen* to quickly switch your attention to how pleasant the steamy shower feels. When you sat down to drink tea, you could have *chosen* to pay attention to how good the tea tasted as you wondered if you might actually prefer tea to coffee. If the same news reporter were to have burst in and asked you how you felt, you would probably have replied, "Great!"

The ability to actively switch your focus from something your mind is drawn toward to something you want to focus on instead feels difficult at first but gets easier with practice.

Your brain is recording a continuous newsreel of your life. It relies on your memories, thoughts, and perceptual experiences to create a representation of the life you live. The thoughts that you permit to enter your mind shape this representation and mold your impression of your life. If your mind assiduously fills itself with happy thoughts, your life will feel happy.

The soul becomes dyed with the color of its thoughts.
                                                —*Marcus Aurelius*

If you have just had a meeting with ten clients and four of them were not smiling, focusing on the six who were smiling will give you a positive memory of the event, whereas focusing on the four who were not will make you think the meeting was a disaster. If you add up five such meetings over the week, you will either feel marvelously successful or hopelessly distraught when the week is over.

- *Consciously train yourself to focus your attention on positive features of your world when you feel tempted to focus on negative ones.*
- *Practice becoming aware of your tendency to choose a negative thought over a neutral one.*

Prolong Attentional Focus

Technology has made life so easy that things that used to require prolonged attention and concentration in the past can now be done quickly with little effort. This is terrible for training your attention skills. Multitasking prevents you from staying focused on one train of thought for long, and you may start relying on your emotional instincts and reward pathways to guide your decisions. Give yourself technology-free time whenever you can.

- *Set aside an hour each evening to make your home a "rational brain cave." Switch off your phone, move away from your computer, play some soothing classical music, and read a good book. Stick with it for at least an hour, even if your mind keeps wandering.*

Self-Control Training

There is a theory that self-control is like a muscle. If you use it intensely or for too long, it becomes fatigued. It can, however, also be trained to work for longer stretches without losing steam. On the basis of this theory, doing something every day that "exercises" self-control may lead to greater success at other tasks that demand self-control. In one study, a group of college students were made to spend two weeks exercising self-control by regulating their posture, their eating habits, and how they felt. Their self-control relating to something entirely different (a handgrip test) had improved at the end of the two-week period.[11] Consciously declining chocolate on a daily basis can lead to greater success when giving up smoking![12]

If this theory is correct, then you have an opportunity to train your self-control every time you make a decision. If you look carefully, you will find plenty of crossroads during a typical day. Many will involve food. Others may involve indulgences. Some will tempt your inner lazy person.

■ *Resist temptation by exercising self-control at every opportunity.*

Consider a situation where you are having a couple of drinks with some friends from work on a Friday night. You are on your third drink and you check your watch; it is late. You were going to get an early night in before hitting the gym tomorrow morning at six, but your friends are pushing you to stay on for a fourth pint. You need to resist the temptation to stay. If you are lifting weights at the gym and had intended to do twenty reps, no matter how tempted you are to stop at eighteen reps, do two more. This won't just benefit your biceps—it will also train your self-control "muscle."

If you only make five decisions like this in a day, you are giving your self-control five training sessions. That's almost two thousand training sessions in a year.

Your heart's adaptability to a new situation, its heart rate variability (HRV), can broadly reflect prefrontal cortical control over other regions of the brain. Higher readings may be associated with emotional regulation and self-control.[13] [14] In line with this, some HRV parameters are seen to rise when someone is resisting temptation. In one study, a group of university students was asked to choose between a plate of carrots and a plate of mouthwatering cookies. As the students exercised self-control and fought against their desire to eat the cookies, some of their HRV parameters went right up![15]

Self-Regulation Training

You are guided by an intention at all times. It may be small (reaching the end of this sentence) or it may be large (meeting a 5 P.M. deadline at work today). This intention is held inside your head like a carrot. The process of *self-regulation* keeps you on track to get the carrot.[16] Self-regulation allows you to "get the job done," whatever it is that you are doing. The process of self-regulation involves self-control, rational appraisal, and decision making. Importantly, self-regulation requires a great deal of emotional regulation. If you are good at self-regulation, you are good at restraining your emotional reactivity.

As in the case of self-control, there is a theory that it may be possible to increase self-regulation capacity through practice. The dlPFC is thought to be one of the key areas in the rational brain to be involved in self-regulation. Musicians rely on a great deal of self-regulation. One study has shown a link between musical training and the size of the dlPFC.[17]

There are two kinds of rewards. Immediate rewards require no patience, little effort, no planning, and no restraint. They are "lazy" rewards. You push a button and get a piece of candy or you place a chip on a green felt table and win a thousand dollars. Delayed-gratification rewards demand self-regulation. So, opting for delayed gratification over instant gratification trains your self-regulatory skills.

Taking on a challenge that requires discipline, patience, and emotional regulation is an excellent way to train self-regulation. Learning a new musical instrument, a new skill, a new language, or a new sport are good examples. It may be possible to improve self-regulation by training other rational-brain-related skills. One unique study has shown that improving working memory through practice might also improve self-regulation in the context of eating.[18] It may also be possible to improve self-regulatory abilities by practicing self-regulation exercises on your smartphone.[19]

- *Choose delayed gratification over instant gratification at every opportunity.*
- *Take up a project that requires grit to complete—and complete it.*
- *Always have a project in the background that you are working to complete.*

There is a well-recognized phenomenon known as mathematics anxiety. People with mathematics anxiety are not able to regulate their negative emotions effectively while solving math problems, and they perform badly in math exams as a result. Brain scans have revealed that the emotional brain of individuals with mathematics anxiety becomes excessively activated when they face a math problem.[20] [21] [22] Helping people regain control over emotional regulation can help mitigate this anxiety.[23]

Two Marshmallows Are Better Than One

A landmark study known as the Stanford Marshmallow Experiment was conducted at Stanford as a series of experiments in the 1960s. In it, a group of nursery school children were told they could either have one marshmallow right now or they could wait a little while and get two. As the children were followed over the years, those who had chosen to wait appeared to have higher SAT scores, better educational performance, and were more successful than those who preferred the instant gratification.[24]

Cognitive Training

Regularly putting yourself in situations that demand a variety of cognitive skills may improve your ability to regulate emotion. It is possible that improving your skill in one area (such as working memory) might improve your skill in another area (such as attentional control). One study has shown, for instance, that people who are pessimistic need greater horsepower to disengage from a negative thought and engage with a positive one, and having a strong working memory can provide that horsepower.[25] It may also be the case that you are inadvertently using attentional focus, self-control, and self-regulation in situations requiring various cognitive skills and these are improved with practice.

Games that challenge working memory have shown good results on emotional regulation.[26] In one study, volunteers were made to learn the sequence of a chain of letters, memorize the order of a series of animals, and remember a range of locations, for twenty days in a row. At the end of the twenty-day period, both their working memory and their ability to regulate emotions had improved.[27]

In 1985, Japanese designers Shigeru Miyamoto and Takashi Tezuka created the platform video game *Super Mario Bros.*, known in Japan as "Supa Mario Burazazu." Twenty-three volunteers with an average age of twenty-four played the 3-D game *Super Mario 64* for thirty minutes a day for two months. At the end of the two months, the right dlPFC of the players had grown in volume.[28] We don't know if this translated into better emotional regulation, but the right dlPFC is thought to be involved in regulating negative emotions through cognitive reappraisal and attentional control.[29]

- *Play a variety of computer games:*
  - *Ones that test your working memory*
  - *Ones that require you to think*
  - *Ones that test your spatial and navigational skills, such as* Tetris
  - *Ones that test observation*
  - *Ones where you need to use many skills at once*
  - *Do this every day for at least fifteen minutes.*

Neurofeedback

A new wave of video games is emerging that use the technique of "neurofeedback." While you play these games, you are wired with sensors. Your brain's electrical activity, your pulse rate, blood pressure, and skin moisture may all be monitored to measure your emotional reactivity and stress at any particular moment. The games involve wild adventures and difficult challenges.

But there's a catch. As you navigate your way through a challenge, your ability to navigate is made weaker if your stress pathways are inappropriately activated. For example, if you are driving a race car, it becomes harder to control the car if you become anxious—your car swerves out of control. The only way you can get it back under control is by getting *yourself* back under control. You must control your stress response and calm yourself down. This is where the genius of the games lies. Most of us are not self-aware enough to know what to do to calm ourselves down. We don't know how to *consciously* activate the brain circuits that carry out emotional regulation.

So, we learn. Through feedback. Initially, we clumsily try anything and everything to somehow reduce the stress response and become calmer. We can't describe in words what exactly it is we are doing because although we think we are doing *something*, that *thing* is happening right at the interface of conscious thought and subconscious implementation. Most of these blind strategies don't work. We know they don't because in the game, the car still swerves. We might have to play around with techniques involving breath control, focus, and seeing things in a different light. Suddenly we might touch on a thought or action that brings the car back under control. Several attempts later, with practice, we can do it on command.

Playing a neurofeedback game of this kind gives you positive feedback on what works (you regain control of the car) and negative feedback on what doesn't (you crash your car) to bring your own heart rate and breathing rate down by consciously regulating your emotional reactivity.

You learn how to control your mind, even if you are not entirely sure exactly how you're doing it. Your brain explores multiple pathways; eventually, it hits on the correct one. Once you do it correctly, repeating it reinforces the circuit. Eventually you *know* what to do to regulate your emotions and control your stress response and can now calm yourself down when you are faced with stress in the outside world.

Another aspect of neurofeedback is *timing*. A stress response is much more difficult to appease *once it is already in full swing*. It is easier to snuff out the fire when it is a small flame and not a roaring inferno. During a neurofeedback game, you learn to identify the earliest signs of your own stress response. Learning what to do to calm yourself down as well as learning *when* to intervene makes a stress strategy even more effective.

A similar effect takes place when we practice yoga.

Yoga

The practice of yoga demands attentional focus and self-regulation, and careful analysis of what happens during the practice of yoga reveals a strong parallel with neurofeedback training. There is emerging evidence that yoga may help to correct an imbalance between sympathetic and parasympathetic activity. Some studies suggest yoga may have a positive effect on depression, anxiety, and post-traumatic stress disorder (PTSD). This may be because in addition to being a workout for your body, yoga provides an excellent workout for your self-control.

Yoga provides an excellent workout for your self-control.

You have miniscule sensors scattered around your neck, chest, and heart that detect variations in blood pressure. These sensors are triggered as you bend your neck into various positions and place your head above and below your heart, and they may signal to your brain to change sympathetic or parasympathetic tone. For instance, if there is an increase in mechanical pressure in the region of your neck, the pressure sensors located there may "think" your blood pressure is high, and they may instigate a rise in parasympathetic tone to bring your blood pressure down. The isometric exercise techniques in yoga are likely to raise sympathetic tone. You also have some receptors in your inner ear that arouse sympathetic activity. These may be activated by changing the position of the head.[30]

As you carry out a series of yoga postures, you are bending forward and backward, "tricking" and triggering your sensors that may raise and lower sympathetic and parasympathetic tone, one after the other.[31] In the midst of all this, you must keep still and focus. Traditional hatha yoga requires that postures be held in complete stillness for a set length of time. Often, you might have to balance on one leg. While your autonomic nervous system is seemingly having a party, your cortex must assert control. This pushes you into a form of focused attention meditation and self-control.

It is only possible to hold a posture if you have control over yourself. If you lose control and become anxious or stressed you will fall out. This provides you with a form of neurofeedback. As in the setting of the neurofeedback computerized games, you will initially not be skilled at controlling your emotional reactivity and modulating your stress response. With repeated effort you will learn what works. Your strategy might incorporate elements of switching focus, controlling breathing, and trying to view things more rationally. Just like in the neurofeedback-based computer games, with practice, the top-down, regulatory circuitry you use becomes stronger and you can apply the skill in your day-to-day life.

HRV correlates with self-regulation—and is seen to rise after practicing hatha yoga. The release from the stretch and muscle contraction (which form a part of every posture) causes a rebound rise in your parasympathetic tone, which reinforces the top-down regulation of your stress response. This sequence of events helps you feel relaxed after each posture. The studies that have shown beneficial effects on HRV have often used traditional hatha yoga practice, which involves static holds rather than dynamic or "vinyasa" postures or novel yoga styles.[32]

To create an even better workout to train prefrontal control, you could choose postures that are uncomfortable and require more effort to focus. Hot yoga may offer such an opportunity. Although we have no empirical information on hot yoga and emotional regulation, balancing in the heat makes it extra challenging to concentrate, and your brain is likely to have to work even harder to calm you down.

■ *Incorporate a no-frills, traditional hatha yoga practice into your daily routine. If you are limited for time, choose only a couple of postures to do every day that require you to balance and concentrate.*

Cognitive Appraisal

There is a saying, "If you change the way you look at things, the things you look at change." When you're in the midst of a stressful experience, your emotional brain is given the stage and your rational brain momentarily relinquishes fine control over your cognitive processes. You are using your emotional brain to assess the situation. Its job is to "predict" danger and buy you enough time, so it hastily scans the scene with a confirmation bias toward negativity. Its perfunctory assessment picks up emotionally contentious signals and fails to spot more complex cues, so it gives you a distorted picture of reality. If you rely solely on this assessment, you will keep your emotional brain unnecessarily activated by making erroneous assumptions. Your confirmation bias intensifies your stress response. The emotional brain has a flaw—instead of forming a theory *after* looking at all the facts, it jumps to a theory shaped by its negative biases and changes the "facts" to fit its theory, adding further fuel to the fire.

Cognitive reappraisal is when you take a second glance at your situation after you have disengaged your emotional brain. You reexamine the evidence rationally and pay attention to subtle features you might have missed before. In so doing, you gather enough evidence to interpret it in a way that causes you the least amount of distress. With practice, you can teach yourself to "read" a situation differently and reduce its traumatic volume.

FINDING THE PIN

In his work *Propos sur le bonheur*, the French philosopher Émile-Auguste Chartier, known simply as "Alain," writes how in any difficult situation, it is important to "find the pin."[33] When a baby cries, its nurse might assume the baby is crying for the sake of crying and not because the pin that is fastened to its diaper has come open and is causing it pain. The nurse grows increasingly frustrated until she finds the pin. Finding the pin (through cognitive reappraisal) can change how you perceive a situation for the better. You won't be too hurt by someone's curt behavior toward you if you happen to know he has just lost his job and is venting his anger. By placing the pin at the epicenter of a stressful event, you remove your "self" as its victim.

USE A THIRD-PERSON PERSPECTIVE

There is some evidence that shifting your perspective from the first to the third person when you recall a negative memory can reduce its negative impact.[34] Thinking about an event as if you were *observing* it rather than *in it* may help you see things clearly without being weighed down by emotions. Free of your reflexive anger, you might empathize more with the person causing you to become angry and see a different perspective that "justifies" your perpetrator's actions. You might spot "the pin."

ASKING *HOW?* AND *WHY?*

Another strategy for effective cognitive reappraisal is knowing when to ask "how?" and when to ask "why?" The question "why?" is philosophical and ruminative, whereas the question "how?" is empirical and prompts fact-based thinking.

If you achieve success, it is okay to ask "why?"
If you fail, it is better to ask "how?"

If you achieve success, ask "why?"
The question "why?" pushes you into existentialism. You think like a philosopher, you ponder, ruminate, imagine, and suppose. Although negative rumination sustains negativity, positive rumination can prolong your feelings of pleasure. You dwell on happy thoughts and this sustains your good mood.[35]

If you fail, ask "how?"
Asking "how?" engages your rational brain as you are forced to carry out a fact-based analysis of the event. Facts matter and your "opinion" doesn't count, so every time your emotional brain tries to pull you away on a tangent, you re-engage your rationality to examine facts.

Here is an example. Imagine you have just lost a cycling race because you recently suffered from the flu and had not fully recovered. See what happens when you ask *why* you failed and compare it to your thought process when you ask *how* you failed.

Q= "Why?"
A= I am a bad cyclist. / I am a failure. / I am weak. / I am unlucky. / I never win races.

Q="How?"
A= The race felt more challenging than the practice run. I became breathless before the uphill section, which is unusual for me. From that point onward I could not catch up. I noticed a drop in stamina after I recovered from the flu last week. My stamina probably needs some more work to return to normal.

Asking "how?" minimizes the negative impact of failure and prevents you from sinking into self-deprecating negativity. It also urges you to identify a way forward and burrow out of the situation. **Active coping**—by taking action in response to a problem rather than passively surrendering to it—is a key strategy for stress resilience.

Asking "how?" minimizes the negative impact of failure.

If you walk away from a stressful experience without having rationalized it in some way, you might find random thoughts about it drifting into your mind and triggering negative emotions at unexpected and often unwelcome moments for a long time afterward. They are your brain's attempts to "work on the problem." In one study, two groups of university students were deliberately put into a negative mood by being made to take difficult tests. They were then asked to write down what they were thinking. One group was encouraged to take a "why" approach, as in "why did this happen?" The second group was told to take a "how" approach, as in "how did this happen?" The "how" approach led to rational reasoning. After twelve hours, the "why" writers felt worse and experienced more intrusive thoughts relating to their experience than the "how" writers.[36] [37]

WRITE IT DOWN

Writing is a form of cognitive reappraisal therapy. If your mind is overwhelmed with negative emotions relating to a past event, writing down a factual account of what occurred forces you to engage your rationality and disengage from your negative emotions. This process of writing has been shown to help relieve social anxiety.[38]

- *Give the story a positive spin in your head—give the least distressing explanation the benefit of the doubt.*
- *Find the pin.*
- *Reappraise the situation as if you were an observer with visualization techniques.*
- *Know when to ask how and when to ask why.*
- *Write down a factual account of what happened.*

Binaural Beats

Binaural beats are being touted as a revolutionary new method of "brain hacking." There is an emerging trend of using binaural beats to become calm as music that sends you to sleep or puts you into a meditative trance becomes increasingly available. Research in support of binaural beats is still in its infancy; however, what is known deserves more than a passing glance.

When information travels along networks of brain cells and astrocytes, it is carried along "waves" of synchronously oscillating electrical signals. When the prefrontal cortex exerts control over networks distributed across the brain, it is thought to do so by acting as a sort of "conductor" of the orchestra of their oscillating rhythms.[39] Cognitive processes such as attention allocation and decision making can fluctuate rhythmically, just like the underlying nerve networks that give rise to them.[40]

Electrical signals passing through brain cells and astrocytes can oscillate in synchrony at several frequencies. Some of these frequencies are:

- Alpha 8–13 Hz
- Beta 15–30 Hz
- Gamma 31–100 Hz
- Delta 0.1–3.5 Hz
- Theta 4–7 Hz

Some interesting patterns of oscillations have been noted in the setting of emotional regulation:[41] [42] [43]

1. As the rational brain regulates emotional reactivity following a stressful encounter, theta waves (4 Hz) may be generated within its networks. The stronger the theta wave power, the better the emotional regulation.[44]
2. Theta frequency activity may be involved in paying attention and during deep meditation.

3. Shifting attention may involve theta and gamma oscillatory electrical activity in the region of rational brain networks.

The entire brain rarely generates the same wave pattern—we
might see one wave pattern emerging here and another one over
there and they spread across relevant regions.

It may be possible to make regions of the brain oscillate
at a chosen frequency by placing electrodes on the scalp to create an electric current or magnetic field (known as transcranial direct or alternating current stimulation or transcranial
magnetic stimulation TMS/tDCS/tACS).[45] These electrodes can
change behavior. For example, inducing alpha brain waves can
help with resisting distractions during a task.[46] It may also
be possible to generate electrical oscillations within the brain
(a process known as **entrainment**) by listening to binaural
beats.

A binaural beat is perceived when you listen to two sine waves
at two different frequencies that are not very far apart, simultaneously, through separate ears. The frequencies must be less
than 1000 Hz. Entrainment is best achieved with a baseline or
"carrier" frequency of between 300 and 600 Hz, with 450 Hz to
500Hz possibly having a more powerful effect. One ear must not
be able to hear what is going into the other ear. We perceive a
binaural beat to be coming from somewhere inside the head and
"detect" a tone at the frequency of the difference between the
two tones. Listening to a tone at 450 Hz in one ear and 454 Hz in
another will result in the perception of a tone at the theta frequency of 4 Hz.

The intention with binaural beats is to induce electrical oscillations at the frequency of the binaural beat. You want your brain

to "dance" to the rhythm of the binaural beat. Unfortunately, this does not always happen. Although listening to binaural beats may change the phase of existing network oscillations, it is not always possible to predict how and where. One study has shown that when experienced meditators listened to binaural beats at the theta frequency, that beat generated delta (and not theta) oscillations (less than 3.5 Hz) in the sides of the brain. When novice meditators listened to these theta wave binaural beats, there was an increase in gamma waves.[47] Although EEG results have been variable, experiments looking at behavioral outcomes show more promise.[48]

- Listening to delta frequency binaural beats for half an hour every day for sixty days brought about a statistically significant improvement in patients with mild anxiety disorders, as measured by an index known as the State-Trait Anxiety Inventory (STAI).[49] In another study on preoperative anxiety, the STAI scores were reduced by 26.3 percent after a session of listening to delta wave binaural beats.[50]
- Theta wave binaural beats may expedite the return to a relaxed state after acute exercise and improve parasympathetic tone while reducing sympathetic tone.[51]

The world of binaural beats is still developing and we still don't fully understand how they work. They are, however, unlikely to cause more harm than listening to other forms of music, and you might find that you enjoy them.

- *Listening to delta frequency binaural beats may help with emotional regulation, although more studies are needed. You*

*might wish to play around with other frequencies of binaural
beats to see if others work better for you.*

Matcha

There may be another way to help your rational brain regulate
emotion. You might consider making it a cup of tea.

Matcha is a specific type of powdered green tea produced only
in Japan. It is different from other green teas in that it is grown in a
way that maximizes levels of the amino acid L-theanine within
the tea leaves. When you drink matcha you are ingesting the
ground leaves, so you are getting a good dose of L-theanine. Like
all tea leaves, matcha also contains caffeine.

Theanine has been shown to improve emotional regulation
and reduce psychosocial stress in mice.[52] It may also have a positive effect on depression in humans.[53] [54] Drinking an extract of
matcha appears to specifically activate the dlPFC.[55]

- *Drink matcha.*
- *Replace coffee with matcha.*

For extra credit:
*Matcha may improve concentration, so if you practice focused
attention meditation after drinking matcha you may find the experience to be more effective. Japanese monks drink matcha before
meditating.*

Diplomacy through Matcha

Matcha has a particular role in diplomacy as demonstrated by Dr. Genshitsu Sen, the XV Grand Master of the

Urasenke Tradition of Tea, with his slogan "Peacefulness through a bowl of tea."[56] Dr. Sen has campaigned tirelessly to promote world peace since the aftermath of the Second World War and uses a tea ceremony as the centerpiece of his arsenal.

If you begin each meeting and every discussion by enjoying a bowl of matcha in a group tea ceremony, several things are put in place. First, drinking tea together is like an initiation into a tribe. Drinking together is an acknowledgment that everyone is part of the same tribe. This reduces stress by putting you at ease. Second, waiting for the water to boil and listening to it on the stove is a form of focused attention meditation. You are activating your rational brain and any anxiety, fear, or anger you may be feeling falls under your control. Third, offering a bowl of the tea to another person is an act of giving that feels good (as you will see later) and also convinces you that the person you are with is a welcome friend and you feel less threatened. Fourth, the circular bowl symbolizes the Earth while the matcha inside represents nature. Dr. Sen says, "I hope that you will feel thankful for the greenery of the planet and that this will lead to peace and an appreciation of nature." Feeling gratitude is calming. Fifth, drinking an extract of matcha has been shown to promote activity in the dlPFC, which may correlate with improved emotional regulation.

By the time this ceremony is completed, a group of anxious, hostile diplomats are at their calmest and most peaceful, so the discussion that follows will be rational, thoughtful, and respectful. You can see why Dr. Sen was designated a UNESCO Goodwill Ambassador in 2012.

His logic applies beyond diplomacy, to relationships, friendships, and in the workplace. If you have difficult meetings ahead with clients whom you wish to appease, drinking tea in this way may make your encounter more peaceful and constructive.

CHAPTER 3

Getting Runaway
Cortisol Under Control

WHEN YOU EXPERIENCE psychosocial stress, your brain shifts gears to set a whole array of chain reactions in motion. These mobilize your brain's networks and prepare your body for confrontation. Your autonomic nervous system is activated and sympathetic discharge increases across your body. Within the brain, a region known as the **locus coeruleus** changes its firing pattern, leading to the release of norepinephrine (noradrenaline). It projects to diverse parts of the brain, including the prefrontal cortex and the amygdala, and influences your cognitive functions and emotional reactivity. Your hypothalamus dispatches a messenger known as corticotropin-releasing hormone (CRH), which prompts your pituitary to send out adrenocorticotropic hormone (ACTH), which then stimulates your **adrenal glands** to release **cortisol**. This trio of organs is known as the **HPA axis**. These processes all feed into each other at various levels.

Different types of stress can affect the autonomic and HPA axis responses slightly differently. One randomized, controlled experiment compared the relative effects of pure psychosocial stress, physical exercise stress, cold-exposure stress, and mental stress from a puzzle on twenty healthy young men. Psychosocial stress caused the largest HPA axis response (as measured by cortisol levels), while physical exercise triggered the largest autonomic response.[1]

All the neurotransmitters and hormones released during an acute stress response act in a coordinated and synergistic way to enable you to respond optimally and to recover as quickly as possible. The effect of one hormone is heavily influenced by the presence of others. Together, they act like a cocktail. All its ingredients are equally strategic and vital for your brain's performance under duress.

The cocktail works just like chocolate: it's amazingly good in small doses, but too much of it makes you feel sick. Imagine you're standing in front of a vending machine that releases chocolates. Each time you feel stressed, the machine releases one. You nibble on it. It makes you smile and feel good again. Suppose that you are now made to feel stressed every couple of minutes. The machine keeps up with its chocolate supply. You must eat the next chocolate before you have completely swallowed the last one. You end up swallowing the chocolates without chewing them. You start feeling rather ill. As your stress triggers become even more frequent, your vending machine is put under strain. It is made to release chocolate after chocolate in quick succession. Eventually it malfunctions.

Too much stress puts you in chocolate overload. Each ingredient in that cocktail contributes to the overload. If you think of stress in terms of the vending machine, then stress becomes a problem if:

1. You frequently encounter it.
2. You take too long to turn off stress hormone production after your stressful moment has passed.

3. You don't *adapt* or you overreact to situations that do not warrant a stress response.

The first time your boss calls you into his office may be a nervous affair. The second time it happens, your stress response will not be quite as intense. This "getting used to new stress" is a vital trait that reduces your day-to-day stress load. If you feel *exactly the same* fear whenever you walk into your boss's office, and you are called in almost every week, your stress load will become rather heavy.

Excessive activation of the HPA axis contributes to the damaging cocktail. Cortisol levels in the blood and saliva reflect HPA axis activity so you can measure cortisol and use it to estimate the course of a psychosocial stress response. If cortisol is a proxy for HPA axis activation in a stress response, then:

- You want to *stop* secreting cortisol when your stressful encounter is over.
- You want to produce *the right amount* of cortisol when you are stressed. Not any more, not any less.
- You want to *adapt* to stress

© HOW DO YOU TURN OFF THE VENDING MACHINE?

Don't Dwell

You walk around with a permanent film reel in your head. This reel is your memory. If you play back a stressful event in your mind just seconds after it is over, and your emotional regulation is defective, then your recollection of the event may activate your emotional brain so powerfully that it perpetuates the stress response. Your body will continue to pump out stress hormones, even though you are simply *thinking* about it and not actually experiencing it.

Imagine that you encounter unexpected traffic on your way to work. After your car has been stationary for a little while, your psychological fear of being late for work triggers a stress response. The effect of this stress response on your body is *similar* to running up a few flights of stairs. Yet, running up a flight of stairs is supposed to be good for you, whereas you know your response in the car is not. The difference between running up the stairs and sitting in the car lies not in what happens *during* the event, but what happens *immediately afterward*. This difference comes from your emotional brain being over-activated.

When you run up a few flights of stairs, your emotional brain is uninterested and does not perpetuate the stress response. The moment you reach the top, you will have no reason to play your running-up over and over again in your mind. As you sit in your car, your emotional brain fills your mind with negative thoughts, fears, and memories. If you have weak emotional regulation, you turn these into a fantastical story line in your head, which agitates your emotional brain into a state of hysteria. You have worked yourself up to such a state that even after the traffic starts moving your mind continues to dwell on just how terrifying your ordeal *could* have been. Notice that your mind has now shifted beyond recognizing facts into the realm of *imagination*. If your emotional brain is activated enough, it may keep pushing on your "stress switch."

If you have a big stress reaction to a situation but you don't dwell on it afterward, its effect on you will be small. If you have a minuscule stress response but you dwell on the event for a long time afterward with compromised emotional regulation, you may continue to activate your HPA axis and its impact will be huge.

This has been demonstrated in the setting of depression.[2] The overall stressful experience of an event is the *sum total* of what happens *during* and *after* it. You see how simply *dwelling* on something however small can make it stressful. This process of "dwelling" is also called "rumination."[3]

Rumination is not harmful in itself, but if you have poor control over your emotional reactivity, then rumination can lead to excessive emotional activation. If your life is stressful and your rational brain is not orchestrating your brain's networks as well as it perhaps should, then avoiding rumination can bring benefits. The tendency to ruminate can predict a new onset of depression.[4] Learning not to ruminate can *prevent* depression and is being explored as a strategy for therapy for depression.[5]

As you sit in your car and drive away after the congestion has eased, you need to summon all available willpower and shift your attention *away* from what just happened and onto another target. If you can rein in your thoughts, you will prevent your mind from wandering into negativity and you will be less likely to sustain your stress response. This is one reason why the *worst* thing you can do after a stressful situation is to relax. Do anything to stay busy and in the present, but don't just relax!

> The worst thing you can do after a stressful situation is to relax.

In a study comparing different types of stress, two groups were given a mental arithmetic test to complete. Not only were they stressed by the test, but just in case anyone was cool enough to sail through the tests without breaking into a sweat, they were harassed during the test, too. Both groups found the overall experience to be emotionally stressful. *Following* this experience, one group was given a quiet and comfortable environment to relax in, while the other group was given something demanding to do, requiring their full attention.

The blood pressure of participants recovered much faster in the group that did not get a chance to relax. In contrast, the first group, having nothing else to focus their attention on, dwelled on what had just happened and played the situation over and over again in their minds. As a result, their stress response remained activated for longer.[6] Other studies have shown similar findings.[7] [8] Insignificant things can become magnified into insurmountable, stressful scenarios, just because you play the situation repeatedly in your mind and have weak emotional control. After a stressful experience, doing something distracting that captivates your attention is much better for your mental health than simply flopping in an armchair to relax. The proverb "an empty mind is the devil's workshop" has more truth to it than we knew!

Dwelling on something, or "ruminating," is also particularly bad at preventing you from getting used to a frequent stressor. If you are new at work, you might at first get palpitations every time your boss walks into the room. After a while, you will get used to your boss and your palpitations will stop. If you ruminate and create an exaggerated story in your head associating your boss with the sensation of fear, you may never get used to your boss.[9]

Disengage Your Emotional Brain

You can use the strategies I described earlier in chapter 2, under "Emotional Regulation," to disengage your emotional brain immediately after a stressful experience and place your full attention on something else. You need to tightly engage your rational brain by forcing yourself to use its functions such as working memory and analytical skills.

Playing computerized games on your smartphone offers a feasible option for doing this in a work setting. Your game must not be too easy, nor so difficult that you are not motivated. Your mind should not have the chance to wander while you play. In one small randomized study, eleven healthy volunteers successfully suppressed negative emotions triggered by recalling sad memories and listening to sad music, either by playing the computer game *Tetris* or by playing a working memory game.[10] *Tetris* is even able to help assuage cravings when one is on a diet![11]

Exercise

Another way to expedite a return to baseline after a stressful experience is light exercise.

"VO2 max" is the maximum volume of oxygen that you are able to use in milliliters per minute, given your body weight. It is measured in ml/kg/min. Exercise can be measured as a percentage of VO2 max. In one study, a group of healthy men exercised for half an hour on an exercise bicycle at 40 percent, 60 percent, and 80 percent VO2 max.

- Exercising at 40 percent VO2 max *lowered* cortisol levels.
- Exercising at 60 percent and 80 percent VO2 max *raised* cortisol levels.[12]

Although there may be a great deal of individual variability, the study's result suggests that gentle exercise at or below 40 percent

VO2 max may lower your cortisol levels. If you are at work, then simply taking a brisk walk around the block is a good idea. Walking or running to a musical beat feels good.

Exercising regularly can influence the effects of ruminating after a stressful experience, perhaps by improving prefrontal control over the HPA-axis response. In one study, two groups of middle-aged women were made to go through a stressful experience while their cortisol levels were measured. One group was sedentary and the other active. Sedentary women with an increased tendency to ruminate experienced a faster cortisol spike and delayed recovery from stress than active women with a similar tendency to ruminate. The active women seemed able to turn off their stress response faster *despite* ruminating.[13]

Use the Power of Your Breath

You can use breathing rate and depth to both instantly dampen a stress response and recover faster from it. Sympathetic activity and breathing pattern are linked. Adjusting one may adjust the other.

One study on people with mild hypertension revealed that slowing down the rate of breathing from 16+/–3 beats per minute to 5.5+/–1.8 beats per minute for 15 minutes reduced sympathetic nerve activity.[14] In another study (on patients with chronic heart failure with high sympathetic tone), reducing the rate of breathing from 16.4 ± 3.9 breaths per minute to 6.7 ± 2.8 breaths per minute and breathing in twice the usual volume of air lowered sympathetic activity by a third.[15]

> ■ *If you are otherwise healthy, slowing your breathing rate down for fifteen minutes, aiming for about six to seven breaths per minute, may help to reduce your sympathetic tone.*

Engage Your Senses

You may be able to help your mind return to a relaxed state immediately after a stressful experience by engaging your senses.

LISTEN TO DELTA WAVE OR THETA WAVE BINAURAL BEATS FOR THIRTY MINUTES.

In a randomized, controlled trial on a group of patients preparing for general anesthesia for outpatient surgery, listening to thirty minutes of delta wave frequency binaural beats significantly reduced levels of preoperative anxiety.[16] Listening to binaural beats at the theta frequency may expedite the return to a relaxed state after acute stress by increasing parasympathetic tone and reducing sympathetic tone.[17] You may want to explore both theta and delta frequency binaural beats to see if listening to either one for half an hour restores your relaxation levels.

LISTEN TO SHAMANIC DRUM BEATS OR MEDITATION MUSIC FOR FIFTEEN MINUTES.

In a study on twenty-nine volunteers, listening to either shamanic drumming or meditation music for fifteen minutes reduced the level of cortisol in the blood.[18] The volunteers listened *actively*, while lying down with their eyes closed.

LISTEN TO NATURE.

Listening to an ocean wave recording for just seven minutes brought down the pulse rate and self-reported stress of a group of college students waiting in a waiting room for fifteen minutes, compared to listening to pleasant classical music or sitting in silence, in one small study.[19] Taking a walk down to your local park and spending some time surrounded by greenery at least once a day is a potent stress antidote that costs nothing, is within easy reach, and has an instant effect.[20] There are many nature sound recordings available online to light up your office and turn it into an exotic Indonesian rain forest for when you have to trudge back to your desk after your lunch break is over.

LOOK AT NATURE.

A randomized crossover study has shown people recover faster from an acutely stressful experience if they have been looking at nature beforehand. Looking at nature raises parasympathetic tone, even if you are only looking at a picture.[21] [22]

SMELL LEMONS.

One study has shown how smelling the scent of a lemon for fifteen minutes can reduce heart rate and blood pressure and make you feel calmer.[23] The volunteers in the study actively smelled the scent for thirty seconds, then gave their noses a break for the next thirty seconds, carrying on in this "thirty seconds *on*, thirty seconds *off*" cycle for fifteen minutes.

Five Strategies to Employ Immediately after a Stressful Experience
1. *Don't dwell.*
   Do something intense, tiring, and distracting immediately afterward.
2. *Disengage your emotional brain.*
   Do something that envelops your rationality such as playing a game that involves working memory, spatial rea-

soning, or analysis. Your activity must be interesting and demanding enough for you to be absorbed. Your mind must not wander.

3. *Exercise.*
   - *Exercise at or below 40 percent VO2 max.*
   - *Go for a brisk walk around the block.*
   - *Listen to music while you run or walk to its rhythm.*

4. *Breathe strategically.*

   Slow your breathing rate down for fifteen minutes, aiming for about six to seven breaths per minute.

5. *Use what you hear, see, and smell.*
   - *Listen to delta or theta frequency binaural beats for fifteen minutes.*
   - *Close your eyes and actively listen to shamanic drum beats for fifteen minutes while focusing on the drum beats.*
   - *Listen to ocean waves or other nature sounds for fifteen minutes and then whenever you can for the rest of your day.*
   - *Head to a park.*
   - *Look at pictures of nature.*
   - *Smell fresh lemons in a thirty-seconds on–off cycle for fifteen minutes.*

The Two Sides of Cortisol

A spike in the stress hormone cortisol can, contrary to expectation, leave you better able to concentrate in some contexts. The trick is timing. A group of healthy volunteers were made to sit through a test that challenged their attention. Before they took the test, some of the volunteers were given a capsule containing 10mg of hydrocortisone (an analogue of cortisol). It was given either just before the test or four-and-a-half hours before the test. Those who were given the capsule four-and-a-half hours before the test performed significantly better on the test than those who had received the capsule shortly before.[24]

◎ MAKING THE VENDING MACHINE RELEASE THE RIGHT AMOUNT OF CHOCOLATE, EVERY TIME

Fitness

There is a theory that repeated exposure to a source of stress in small doses reduces the stress response to stress coming from the same or even a different source. This is known as the *cross-stressor hypothesis*.

> *What does not kill you makes you stronger.*
> —*Friedrich Nietzsche*

Exercise is a stressor. As you mount a *small* stress response to exercise repeatedly, your stress response to it becomes smaller. Regular exposure to exercise makes you better at recovering from it.[25 26 27 28] This improvement is not permanent—if you quit exercising, any such improvement is reversed.[29]

According to the cross-stressor hypothesis, with regular exercise, your stress response to an event in an entirely different setting may decline, too. Some studies looking at the relationship of fitness and exercise and psychological stress seem to support this idea.[30] For instance, if you exercise aerobically for thirty minutes, this can reduce your stress response to a stressful situation ninety minutes later. The degree by which it reduces your stress response may be inversely related to how stressed you are while you exercise. Being fit in general helps your state of mind and this might influence how you respond to stress.[31] As you may remember, being fit may protect you from the stress-perpetuating effects of rumination.

The cross-stressor idea is a hypothesis and some studies have not shown clear beneficial effects of exercise stress on stress reactivity. It is clear, however, that improving your general health and fitness will benefit your overall well-being. This is likely to influence how you respond to stress. One study on a cohort of college students has shown how regular exercise can reduce perceived stress and personal burnout.[32] Improving your general health with regular exercise may help you to be resilient in *all* stressful situations.

Improving your general health with regular exercise may help you to be resilient in all stressful situations.

If you are exercising intensely, take long breaks between sessions. The more intense an exercise session, the longer it takes to recover from it.[33] [34] Exercising before your cortisol levels have dropped back to normal after intense exercise will leave you with even higher resting cortisol levels.[35]

- *If you are unfit, focus on improving your fitness.*
- *If you are fit, focus on maintaining your fitness.*

A Helping Mindset

Doing something as simple as changing your mindset can dampen your hormonal stress response to a forthcoming stressor. Imagine you are on the show *The Apprentice.* You have been assigned team leader for a task in an area that you find the most difficult. Your hard-selling ability will be put to the test. You are terrible at selling. You will enter the task (just as everyone else will) with a focus on outcompeting the others, since you want to win. This is where you need to make a switch.

Instead of entering with a mindset intent on elevating yourself above others, switch to a mindset of *assisting* others. We can create "social" stress in a laboratory using a setup known as a Trier Social Stress Test, or TSST. When a group of healthy eighteen- to forty-five-year-olds were made to spend just a few minutes before being tested thinking about their values directed at helping others and how they would like to implement these values through their work, they had a milder hormonal stress response to the test. Their HPA axis activation was reduced and they produced significantly less cortisol, compared to others who did not do this mental prep work.[36]

This strategy may work because needing to prove yourself is a *defensive* stance. It is you *against* them. The need for defense pushes you toward vigilance and stress. Seeing others as part of your own team convinces you that you are part of a tribe—and that you have tribal protection. Here, it is you *with* them. You no longer need to feel defensive. There is no need to be stressed.

One way of implementing this incredible tool is by approaching your day at work focused on the overall journey of your team or of your entire company. You can train your mind to see others as team members rather than competitors by actively helping or looking out for them, almost as if it were a ritual. Even if you are in a cutthroat competitive environment, this approach is likely to give you an edge as it may help you to stay calm and focused in the long run.

- *Train yourself to see others as players on the same team rather than as rivals.*
- *Constantly remind yourself of the aims you share with others, and focus on them.*
- *Create a ritual of helping others, just for the sake of it.*

NUTRITION

Certain aspects of your nutrition can affect your stress response.

Eat natural probiotic yogurt.

A double-blind, placebo-controlled, randomized trial found that eating 100g of probiotic natural yogurt every day for six weeks reduced general perceived anxiety and stress.[37]

Drink water.

Staying well-hydrated may reduce your HPA axis response to stress.[38] Dehydration can affect your mood in a negative way.[39]

Avoid food with refined carbohydrates or added sugar.

One randomized trial found that eating a meal that has a high glycemic index at breakfast time may increase cortisol levels compared to a meal with a low glycemic index.[40]

Get enough salt.

When a group of healthy eighteen- to thirty-eight-year-old men and women were given a sodium-restricted diet for five days, their baseline sympathetic activity increased and their parasympathetic activity decreased. Their resting sympathetic tone was raised to such a degree that it could not climb much higher when they were made to feel stressed. When they returned to a "normal sodium" diet, their baseline sympathetic tone fell back to normal and rose in a healthy way to stress.[41] Unless you have been instructed otherwise by a physician, make sure you get enough salt.

OXYTOCIN

Oxytocin is sometimes called the "love hormone" because its levels are increased in the setting of social bonding. We have a human need to belong. Belonging makes us feel safe.[42]

When rats are shown a ball of fur, they become terrified and respond in one of two ways. Those with higher oxytocin levels reach out to each other and huddle together in defense. Those with lower oxytocin levels are more likely to deal with the situation by themselves.[43] If you take a group of healthy humans and put them through a stressful experience, those who have a best friend with them during the experience will have a smaller spike in cortisol than those who are alone. Giving the volunteers oxytocin reduces this spike even further. Oxytocin *plus* a best friend is better than only having oxytocin or only having a best friend present. Both social support and oxytocin help to suppress stress reactivity.[44]

Oxytocin levels may be raised by feeling loved, having a supportive partner at home, sexual orgasm in both men and women, simple human touch *as long as the touch is occurring within a safe and appropriate context*, and Swedish massage therapy (the effect can last for several days).[45] [46] [47] [48] [49] [50]

In the old days, we would probably head next door to see our favorite aunt if we felt a little down. She would greet us with a warm hug and a cup of tea, and this would make oxytocin levels surge. A cuddle with Grandma used to be a far better painkiller for children after a cut or bruise than taking a pill. Nowadays, aunts and grandmas no longer live next door. They may not even be on the same continent. There is often no one to cuddle with when we get home. Our world has changed so much that there now exists a profession of "cuddling."[51] A professional cuddler can make $60 an hour. There are even rumors of a national cuddling certification course to be launched soon in the United States.

As you can see, you have many options for raising your oxytocin levels. The best option might be Grandma but if you have $60 to spare, you can hire a cuddler, too!

- *Prioritize a loving relationship.*
- *Hug people whom you are close to, at every opportunity.*
- *Get a Swedish massage.*
- *Invest in a good social network and a "tribe"—religious groups, athletic clubs, and social clubs are all good options. Please note that "virtual" groups on the Internet are unlikely to be as effective as "real" groups whose members you see in the flesh.*

SMILE!

Smile . . . with your eyes as well as your mouth.

In one experiment, seventy healthy participants were divided into three groups. One group held a chopstick between their teeth. The second held the chopstick while actively clenching the teeth so as to activate the "zygomaticus major" muscle, which contracts when you smile. The third group was instructed to do the same thing as the second group while contracting the muscles around the eyes (orbicularis oculi), which are activated during a "genuine" smile. The second group recovered faster from a stress response than the first group. The third group recovered faster still. The second and third groups both demonstrated a reduced stress response. *Smiling dampened the response to stress.*[52]

Some studies looking at facial muscle activation and mood have not always replicated these findings; there is, however, little reason not to smile, regardless.

BOOST PERCEIVED CONTROL

If you *think* you are in control of a situation, it may stress you less and you will likely produce less cortisol.[53]

If you expect an upcoming stressful scenario, assess all the aspects of the scenario that fall within your control and then assert control over these aspects. Ahead of a meeting you can have control over the venue, the seating plan, the temperature, and the lighting, and having an exit strategy will give you control over your exposure time. You may also convince yourself that you are in control over the discussion by having contingency plans for as many predicted outcomes as possible. For instance, you may decide to reflexively interrupt each negative downturn with a positive affirmation.

⊚ STAYING RELAXED

You can adjust certain aspects of your lifestyle to maintain a healthy sympathetic and parasympathetic balance and normal cortisol levels on a day-to-day basis.

Exercise—in the Morning

Exercising in the morning as opposed to the evening may raise parasympathetic activity at night while you sleep, improving your sleep quality.[54]

Visit a Sauna

Waon therapy is a Japanese sauna therapy protocol where people sit in a dry infra-red sauna set to 140 °F for fifteen minutes and then leave the sauna and rest, wrapped in a blanket, for thirty minutes. In one study, doing this **once a day, five days a week for four weeks** reduced baseline sympathetic tone and increased para-sympathetic tone.[55] In another study on ten people suffering from chronic fatigue syndrome (CFS), this protocol improved anxiety, depression, and fatigue after the four-week period.[56] Attending a regular sauna for **twenty minutes, three times a week for two months** significantly reduced tension headaches in a group of volunteers in a randomized study.[57]

Avoid Negative, Stressed People

When you look into someone's eyes, you may synchronize your pupils with that person's. Your pupil size reflects your stress level. If you interact with calm people, you are more likely to become calmer. If you surround yourself with genuinely happy and positive people, you "catch" their happiness. Mood is infectious. So avoid negative people!

Mood is infectious.

A study that followed more than four thousand people in a social network over twenty years found that happiness can extend to up to three degrees of separation—to the friend of a friend of a friend. If you are surrounded by happy people, you are more likely to be happy in the future. This effect even extends beyond your immediate surroundings. If you have a friend who lives within a mile of you and that friend becomes happy, the chances of you becoming happy go up by 24 percent.[58]

This effect is magnified by the world of social media. If you scroll through your social media newsfeed and find negative posts, you are much more likely to enter into a negative frame of mind. Curate whose posts you see in your feed, with this in mind.

THREE WAYS TO STAY RELAXED
- *Exercise in the morning.*
- *Regularly visit a sauna.*
- *Avoid negative people, stories, and situations.*

An Emerging Therapy Involving Eye Movements and Bad Memories

A funny thing happens when your eyes track a moving object. The bad thoughts, beaming inside your mind at that moment, may lose their luster. In his first novel, *Une vie*, published in 1883, Guy de Maupassant wrote about a young woman returning to her bedroom in a state of mental anguish when she spotted a bee flying from left to right and right to left in a fast but flowing movement. As she gazed at the moving bee, she suddenly "snapped out" of her negative emotional state. The French neuropsychologist Olivier A. Coubard points out that since this peculiar phenomenon was not known at the time, de Maupassant must have experienced it himself in order to write about it.[59]

In the 1980s, Dr. Francine Shapiro, a psychologist, was walking along a beach when she noticed how her eyes would involuntarily move from side to side whenever a negative thought entered her head. The thought would then disappear. Dr. Shapiro wondered if she could influence a negative thought by voluntarily moving her eyes. As she did so, "the thought shifted away" and when she brought the thought back, "it wasn't as disturbing."[60] Dr. Shapiro decided to investigate further, and her research has led to a revolutionary therapy known as Eye Movement Desensitization and Reprocessing therapy or EMDR. It was first dismissed as a placebo effect, or a form of exposure therapy, but accumulating evidence suggests it may be neither. EMDR therapy is based on the uncanny ability of eye movements to diminish the emotional load of a negative thought.

A therapist moves a finger or hand rapidly, usually from left to right and right to left, at a speed of around 1.12 Hz, for about twenty-four seconds. The client follows the movement carefully with their eyes, without moving their head, while recalling an unpleasant memory. The emotional load and vividness of the memory grow smaller during the EMDR session. As the negative load of the memory is reduced, it becomes easier to associate the memory with new, positive affirmations.

A randomized, controlled trial testing EMDR therapy on adult Syrian refugees from the Kilis refugee camp at the Turkish–Syrian border showed EMDR significantly reduced PTSD and depression symptoms.[61] Another randomized, controlled trial showed its efficacy in the setting of postoperative pain in teenage patients.[62] When a group of active-duty service members affected by post-traumatic stress in the United States were given EMDR therapy, they needed fewer sessions of psychotherapy.[63]

We don't yet understand how EMDR works. One potential route is through working memory, which has a limited capacity. Paying visual attention to something consumes a large part of your working memory reserve. If on top of that, you try to focus on a painful memory, you can't dwell on the entirety of the recalled experience.

During a stressful episode, your memory usually records *what you feel*, rather than what *actually happens*. Experiences that feel particularly traumatic are recorded by your brain in vivid technicolor so they stand out. When something—an image, a scent, or a sound—triggers the memory of this type of experience, the entire memory flashes back with special effects, exaggerated color, and deafening sound. An innocent trigger sets light to an entire network of thoughts, images, and emotions that usurps whatever is occupying your mind at the time. If your mind's resources are limited when the memory is recalled, you may be forced to recall the memory *without* the special effects. Perceiving the memory without its vividness and emotional load may change the original memory and this may force your brain to store it in its new muted form.

YouTube offers many videos of a target that moves from side to side at a frequency of 1 to 1.2 Hz, mimicking the EMDR technique, but psychologists strongly recommend that EMDR be attempted only with a trained therapist because a resurfacing of negative memories and emotions can sometimes cause immense distress. *(Please note: EMDR videos should not replace formal therapy sessions or be used in the setting of a formal diagnosis of post-traumatic stress disorder or other conditions without the approval of your physician.)*

CHAPTER 4

Fostering Growth
in the Rational Brain

A s you read this sentence, your working memory compares what you are reading now with what you read just an instant ago. It records every fleeting moment in a temporary logbook, connects the dots, and creates a narrative of continuity to make time appear unbroken.

Your rational brain analyzes data from your working memory and your long-term memory stores to form assumptions, make predictions, and *upgrade* its existing strategies so you can adapt even better to your world. As it does this, coherent electrical oscillations ripple across networks of brain cells and astrocytes. Aggregates of brain cells dance together in rhythm, and one brain cell may dance to several different rhythms. This vibrant scene is filled with vigorous synaptic plasticity as synapses within networks strengthen and weaken. The cells in your brain produce chemicals that encourage growth. One such chemical is brain-derived neurotrophic factor, or BDNF.

These "growth promoters" nurture dendritic branches, stimulate the creation of new synapses, and help brain cells survive. The synaptic landscape steadily evolves as new synapses form while others are pruned away. Your prefrontal cortex and hippocampus are in a state of *constant growth*.

> *Your prefrontal cortex and hippocampus are in a state of constant growth.*

When you are not stressed, your rational brain carefully choreographs your behavior to lead you toward a purposeful goal. In acute stress, you abandon this goal-directed behavior and become instinctive, as you respond without thinking to what you see in your environment. This renders some of the work carried out by the rational brain unnecessary and obsolete.

As the rational brain adapts to a paradigm of *chronic* stress, its normal pattern of synaptic plasticity and growth is reshaped for efficiency.[1] Long-term occupational stress is associated with shrinkage in areas of the prefrontal cortex, including the dlPFC.[2] Studies on chronically stressed animals have revealed how synapses within parts of the hippocampus and prefrontal cortex wither and reorganize.[3] The dorsal half of the hippocampus, which serves learning and memory functions and is part of our "rational brain" concept, undergoes shrinkage.[4]

In contrast to what happens in the rational brain, there may be an *enhanced* state of growth in parts of the brain that process emotion in some instances of chronic stress. One part of your amygdala, the basolateral amygdala, influences anxiety. Chronic stress can raise BDNF levels in the basolateral amygdala and the dendrites of its brain cells can grow longer and branch more.[5] This growth correlates with anxiety. Intriguingly, while the dorsal hippocampus shrinks with chronic stress, the ventral hippocampus, which forms part of the emotional brain, enlarges.

In light of these findings, there is now a theory that clinical depression may be caused by hampered growth in the rational brain.[6] Current approaches to treating depression include discovering new ways to stimulate its growth. This pattern of brain changes is also seen in aging, prompting researchers to wonder if boosting growth in the rational brain might decelerate the aging process.

If chronic stress skews the pattern of growth away from your rational brain, then doing all you can to *actively encourage* growth within the rational brain may help you to offset the effect of chronic stress.

The following may help to nurture growth in the rational brain:

1. Exercise
2. Heat or cold[7]
3. Environmental enrichment[8]
4. Caloric restriction
5. Turmeric

⊚ EXERCISE

Non-intense exercise increases the birth of new brain cells and astrocytes in the hippocampus in mice and BDNF levels rise within minutes to hours after starting to exercise.[9] [10] [11] Not all types of exercise have the same effect.

Types of Exercise

In one study, a group of rats was given one of three exercising protocols to follow over a seven-week period.[12] (A control group did no exercise at all.)

A. They had a running wheel in their enclosure and could take a jog whenever they felt like it.
B. They were put through weight training exercises (they had to climb a wall with little weights attached to their tails).
C. They were made to train with fifteen minutes of high-intensity interval training (HIIT). They were forced to run very fast on a treadmill for three minutes, rest for two minutes, and repeat three times.

Weight training (option B) had no effect on neurogenesis when compared to the control group. High-intensity training (option C) had a small effect. The most pronounced effect came from voluntary, enjoyable, light jogging (option A). The number of new brain cells the rats generated in the hippocampus correlated with the total distance jogged over the seven-week period. This finding corroborates the results of previous studies that have also shown that mild exercise is better for rational brain growth than intense exercise.[13] The effect also holds true for humans.[14] Exercising at moderate intensity three days a week for one year reversed age-related shrinkage of the hippocampus in older adults by one to two years in one study.[15]

As exercise exerts its beneficial effects on the brain, there is concurrent improvement in memory, learning, and general cognitive abilities.[16] Sleep deprivation reduces BDNF levels. Only four weeks of regular exercise on a treadmill *prevented* memory impairment and a drop in BDNF levels brought on by sleep deprivation in one animal study.[17] [18]

Too Much of a Good Thing: The Downside of Too Much Exercise

Too much exercise acts like *chronic* stress, and intense or excessive exercise can make the hippocampus shrink in volume *even if it improves cardiorespiratory fitness*. In one study, seventeen healthy young men were made to follow a high-intensity physical fitness program involving exercising three times a week for sixty-minute sessions over a six-week period.[19] The exercise intensity was carefully monitored and individually tailored so that everyone exercised at the same intensity relative to their own fitness level. At the end of the six weeks, the young men had markedly improved their fitness scores as shown by a 5 percent rise in their VO2 max. While their VO2 max rose, their hippocampi shrunk in volume by 2 percent on average, across the group. It is difficult to know what caused the loss of volume, but whatever the cause, the shrinkage occurred in association with falling levels of BDNF, which makes it likely that it corresponded with a decline in growth. An inflammatory agent, TNF-α, was markedly elevated at the end of the training program. Intense exercise can lead to inflammation (as you will discover in chapter 6) and high levels of inflammatory agents can damage synapses and hinder learning and memory, which may have contributed to the findings in this case, at least in part.[20] [21]

Since inflammation reduces growth, standard anti-inflammatory medications such as aspirin and Tylenol are currently being investigated as potential agents to enhance growth in the hippocampus.[22]

How much exercise is too much? The *lactate threshold* is a measure used by exercise scientists to gauge exercise intensity. One mouse study has shown that six weeks of exercise at *below* the lactate threshold encouraged neurogenesis in mice, whereas a similar period of exercise *above* the lactate threshold did not.[23] The lactate threshold is different in mice and men, and this effect may not translate into humans, but we can try to estimate what the equivalent exercise threshold is likely to be in humans.

It is best to have your lactate threshold measured by a fitness professional, but a very approximate rule of thumb is that it is the point where you reach 85 to 90 percent of your maximum heart rate. Your maximum heart rate may in turn be roughly estimated by subtracting your age from 220. So, if you are forty years old, your lactate threshold may lie around a heart rate of 160 beats per minute. If we were to extrapolate from what happens in mice, making sure your heart rate stays well below this number may help you avoid excessive inflammation. In high-intensity exercise training the heart rate can climb above this threshold, which may explain the findings from mouse studies where high-intensity exercise training did not enhance neurogenesis quite as expected.

WORKOUT TEMPERATURE

Animal studies suggest it might be possible to stimulate the state of growth in the rational brain with exercise in the ambient heat or cold. Rats who exercised either in the heat (99.5 °F) or in the cold (40.1 °F) had more new brain cells than rats who exercised at room temperature for five days.[24] Long-term exposure to mild heat also enhances neurogenesis in mice. Mice who suffer from severe traumatic brain injury recover faster if they are kept at a temperature of 93.2 °F for four weeks before the injury.[25]

@ A LEARNING ENVIRONMENT

You want to drive your rational brain into a state of high growth and activity, but your intelligent brain is not going to put in the effort to become active without good reason. Like all intelligent creatures, it is lazy and does not want to waste its energy on point-less endeavors. So, you need to tempt your rational brain with a reason to become active by presenting it with a carrot hanging from the end of a short stick.

This carrot takes the form of two things: challenge and nov-elty. You overcome a challenge by using old ideas in a new way, as you analyze information and form a strategy. When you're in a new environment, you must learn how to navigate within it, associating cause with effect and identifying new rules. These behaviors depend on new networks and connections, which re-quire extensive synaptic activity in your prefrontal cortex and hippocampus. An environment that presents both novelty and challenge provides *environment enrichment* (EE), and tempts your rational brain into becoming more active.

A growing number of animal studies indicate that EE can buffer the effects of chronic stress. At present there is insufficient data from human studies on the effect of EE on chronic stress, but early observations suggest we can benefit from EE, too.

In one experiment, rats were put through six hours a day of chronic stress every day for three weeks. The experience im-paired synaptic plasticity in their hippocampi, increased anxiety, and affected working memory. They were then exposed to EE for six hours a day for ten days. The enriched experience completely reversed their anxiety, ameliorated working memory, and improved synaptic plasticity.[26] Another study has shown that rising levels of BDNF may partly contribute to an increase in synaptic plasticity.[27] [28] While EE increases synaptic plasticity in the rational brain, mouse studies reveal it has the *opposite* effect in the emotional brain. Chronic stress can encourage dendritic growth in the basolateral amygdala, and the degree of growth correlates with worsening anxiety levels. There is some evidence that EE reverses this growth and can normalize anxiety levels.[29] If depressed mice are placed in an enriched environment while being treated with antidepressant medications, their medications are more likely to work.

Video games provide an opportunity for EE. In one study, a group of video gamers played either a three-dimensional game (*Super Mario 3D World*) or a two-dimensional game (*Angry Birds*) for thirty minutes every day for ten consecutive days. At the end of the ten-day period, the *Mario* gamers had improved memory and cognitive skills compared to the players of *Angry Birds* and a control group. Not only did their spatial skills improve, but their nonspatial memory did as well.[30] It is difficult to study synaptic plasticity in a human brain, but an improvement in cognitive skills and memory is likely the result of increased activity and synaptic plasticity in the regions of the brain that serve these functions, such as the prefrontal cortex and the hippocampus.

Similarly, musical training offers a form of EE, as you integrate auditory information with visual data while following complex rules. In a study on people aged sixty to eighty-four, musical training over a four-month period improved mood, cognitive skills, attention, and executive function.[31]

If you find your work dull and unstimulating, incorporate an activity into your day that stimulates and challenges you. A recent case-control study on middle-aged production workers who had been doing the same work for the past seventeen years revealed that those who had experienced new things at work more often during that time frame had better cognitive skills.[32] Mihály Csíkszentmihályi's work on *flow* has shown that the *flow* state can only be achieved if you feel challenged, and people who incorporate *flow* into their days tend to suffer less from stress. If your work is uninspiring, consider doing what you usually do slightly differently or give yourself tasks that make your work feel more challenging. When I was training in ophthalmic surgery, after establishing that I was able to carry out a cataract extraction procedure safely, my consultant encouraged me to start practicing the microsurgical procedure using opposite hands. During the course of his own career he had set himself challenge after challenge to become increasingly creative in a constant setting. This exceptional trait led him to eventually pioneer cataract surgery without the use of local anesthesia, after he had trained himself to operate on a moving eye. He remains one of the most fulfilled medical professionals I know.

Welcome new experiences at every opportunity.

Never allow yourself to feel bored. Stimulate your brain in a variety of ways. Welcome new experiences at every opportunity. If you choose to play video games, choose ones that integrate a diverse range of skills.

- Never be bored.
- Stimulate your brain in as diverse a way as possible, every day.

⊚ CALORIC RESTRICTION

Eating fewer calories than the body needs has been shown to increase neurogenesis within the hippocampus in multiple animal experiments.[33] Humans also may benefit from caloric restriction. When fifty healthy older volunteers (with an average age of 60.5 years) restricted their total daily calories by 30 percent for three months, their memory performance improved.[34] [35]

- *You do not need to restrict your calories; simply limit them to no more than what you need. Try to make this a habit for life.*
- Please note: Caloric restriction is different from the condition of anorexia nervosa, in which sufferers become nutritionally deficient. It is vital to practice caloric restriction under the guidance of a physician, so that health is not compromised.

⊚ TURMERIC

Curcumin, an ingredient in the spice turmeric, increases neurogenesis in mice in a dose-dependent way.[36] If you were to take a group of rats and put them through chronic stress, levels of BDNF in certain parts of their brains would fall and they would become depressed. Their mental performance would suffer. In one experiment, giving the rats turmeric reversed these changes.[37] At least two double-blind, controlled, randomized trials have confirmed an antidepressant effect of curcumin on humans.[38] [39] As you will discover later, eating turmeric as a whole food may be more beneficial than taking an extracted, processed supplement.[40]

☺ WHAT TO AVOID

We have identified a few factors that may adversely affect activity in the rational brain.

1. Chronic inflammation

   Inflammation in the rest of the body can affect neurogenesis in the rat brain.[41] [42] We'll look at inflammation in detail in chapter 6.

2. Tobacco smoking

   Some poor mice were made to inhale tobacco smoke for the sake of science. The smoke inhibited neurogenesis in the hippocampus.[43]

3. High blood pressure

   Hypertension has been shown to impede dendritic branching in the hippocampus of mice.[44]

CREATING AN ENVIRONMENT FOR BETTER GROWTH IN
THE RATIONAL BRAIN

- Make a habit of regular, aerobic, treadmill-like exercise with episodes of resistance training.
- Be sure to exercise if you anticipate stress or sleep deprivation in the weeks ahead.
- Don't over-exercise.
- Whatever you do during the day, use the opposite talents when you get home.

- Seek out new experiences to stay stimulated.
- Challenge yourself.
- Keep things around you that "tickle" different senses.
- Enrich your environment to present yourself with new challenges.
- Eat within your caloric limit. Try to make this a habit for life.
- Consider adding turmeric to your cooking (see chapter 7).
- Don't smoke!

CHAPTER 5

Tuning Your Body Clock

I wonder if I've been changed in the night. Let me think.
Was I the same when I got up this morning? I almost think
I can remember feeling a little different.
—Lewis Carroll, *Alice in Wonderland*

ALICE IS RIGHT. She has indeed been changed during the night. You wake up with a slightly different brain to the one you had the night before. Your rational brain choreographs your behavior with expert precision on a moment-by-moment basis while you are awake. This studied choreography generates frantic activity as new networks form and synapses evolve. The longer you remain awake, the longer this state of intense activity and growth persists. Eventually, your brain is overwhelmed. It needs a break. Sleep is the necessary respite—it's like taking your car to the garage after a long trip for a general tune-up, clean-up, and replacement of worn-out parts.

As you sleep, your brain sifts through its inventory. Networks and synapses are refined and consolidated, and redundancy is removed. Some memories may become stronger while others are made weaker. By the time you wake up, your brain's landscape looks different than it looked when you fell asleep.

Sleep also provides a "clean-up" service. The brain is surrounded by an irrigation system called the glymphatic system. It redistributes useful products and clears away "waste." Animal studies have shown that this system is more active during sleep than during wakefulness. Astrocytes are large star-shaped cells that nurture brain cells and their synapses. They are involved in the glymphatic system and they seem to redistribute themselves over synapses if the brain has been kept awake and active for a long time.[1] This may lead to a slight swelling of the brain, which is reversed after a good night's rest.[2]

When you sleep, your brain transitions through several states. In one, your eyes move rapidly from side to side, known as rapid eye movement (REM) sleep. In another, you are in deepest slumber, called slow-wave sleep (SWS). You cycle between these two states and in between various other states. It seems the longer you stay awake, the more desperate your brain becomes for SWS. When you sleep after staying awake for twenty-four hours, your SWS phase will become prolonged. People with depression seem to have a shorter SWS phase than normal, and this is thought to explain why sleep deprivation for a night, by increasing SWS the following night, can temporarily improve the symptoms of depression.[3]

SLEEP AND PTSD

Sleep reduces the impact of a stressful experience. People who sleep well before, during, and after the experience of severe trauma or stress tend to be less affected by mental illness afterward. Those who have disturbed sleep soon after the experience are more likely to develop PTSD, depression, suicidality, and substance abuse problems.[4] Just as it is possible

Sleep reduces the impact of a stressful experience.

to learn to associate a cue with a sense of fear (for example, an ambulance's siren may become associated with fear in someone who has just witnessed a terrifying accident), it is possible to *unlearn* that association. People who bounce back after a traumatic experience, including those who overcome PTSD, are thought to do so partly by *unlearning* the association between reminders of the trauma and memories of the trauma itself. Sleep—particularly REM sleep—is important for this unlearning process.[5]

⊚ FALLING ASLEEP AND WAKING UP

Given that sleep is your brain's tune-up garage, the longer or the more challenging its mileage, the more desperate it grows for routine maintenance. The longer you remain awake or the more mental activity you accumulate, the greater is your brain's need for sleep. You start gathering a *sleep debt* from the moment you wake up in the morning. The debt grows heavier and heavier until you are desperate to resolve it at night. But there's a catch: your body works to a twenty-four-hour clock, and no matter how heavy your sleep debt is, you can't fall asleep if it isn't the correct time on your clock. This clock is like a security guard of punctuality at the entry gate to whichever land you are trying to enter, be it the awoken world or the asleep one. You may start finding your sleep debt growing rather heavy, late in the evening, but your clock will only allow you to enter the land of sleep when the time is right. By around four in the morning, you may have replenished all your sleep debt but you are not permitted to enter the land of wakefulness yet, so your body clock keeps you asleep for another few hours.

If you have a large sleep debt but a malfunctioning clock, you won't be able to sleep. If you have a perfect clock but no sleep debt (from napping in the afternoon) you won't be able to sleep either. For a good night's sleep, you must take care of your body's clock.

@ THE BODY'S CLOCK

You don't have one body clock. You have *thousands!* Heart cells have their own clocks, as do liver cells and intestinal cells. Multiple clocks are even scattered across various parts of your brain. Each has its own rhythm. There is one central clock sitting in the brain in a region called the suprachiasmatic nucleus. Like the master conductor of an orchestra, it keeps all the other clocks, known as peripheral clocks, synchronized, so they all show the same time.

Chronic stress can disturb your body's twenty-four-hour clock, and disrupt its circadian rhythm.[6][7][8][9] On the other hand, if you break your circadian rhythm, you may increase your susceptibility to chronic stress.

Your central clock keeps your peripheral clocks synchronized by sending down neuronal and hormonal signals. One of these signals is thought to be cortisol.[10] If chronic stress makes your cortisol levels go awry, this can affect the alignment of your clocks. A healthy body produces cortisol with a natural rhythm that follows a circadian pattern. There is a small surge in the morning and an ebb during the night. If your circadian rhythm becomes muddled, the normal fluctuation of cortisol levels (and HPA-axis function) is interrupted and this may affect how you respond to stress.

There is a theory that a stressful lifestyle disrupts the body's clock, and this disruption gives rise to depression and other mood disorders.[11] [12] Any activity (such as intense exercise), experience (such as excitement or anger), or food (such as coffee) that changes your normal cortisol rhythm, and that makes your cortisol levels rise when they are not supposed to, may disturb your circadian rhythm.

◎ MELATONIN

Melatonin is sometimes called a "darkness" hormone. It is released by a gland called the pineal gland. In the absence of bright light and blue light, your master clock, the suprachiasmatic nucleus, tells your pineal gland to release more melatonin into your blood. In a way, your master clock talks to the rest of your body through melatonin as it swims around your body telling all the peripheral clocks to march together.[13] Emerging research suggests that melatonin also has anti-inflammatory and antioxidant effects.

Melatonin at night may calm the stress response.

Melatonin at night may calm the stress response. It is currently being explored as an option for treating patients who suffer from high blood pressure during the night, as this may be caused by too high a sympathetic tone.[14] [15] Melatonin can reduce the sympathetic response to mental stress.[16]

Early results from animal studies suggest melatonin might protect the brain from the effects of chronic stress.[17] One study has shown that agomelatine, which mimics the effect of melatonin, reversed many of the effects of chronic, unpredictable, mild stress in male mice, including behavioral alterations and changes in the hippocampus.[18]

It is not simply the presence of melatonin that is important. There is a burgeoning theory that although boosting melatonin during the night may help depression, failing to curb melatonin production in the morning may actually *cause* depression.[19] It seems that melatonin must work *with*, and not *against*, the body's clock to be effective. Now we'll take a look at how to achieve this.

© ADJUSTING YOUR BODY CLOCK

As chronic stress can leave your body clock in disarray, and disrupted sleep and biorhythms add fuel to the fire of chronic stress, doing all you can to maintain a regular body clock will work against the effects of chronic stress.

You can use the information we have gathered so far to tune your body clock using four aspects of your lifestyle: light, food, exercise, and heat. Things that make you excited or alert also affect your clock. In general, light, food, exercise, and heat are things we engage with while we are *awake*.

Light

Light can be manipulated to trigger melatonin production at night, improve its production, and reduce production when you wake up. Remember, it may be as important to reduce its production when you wake up in the morning as it is to increase its production during the night. The receptors in your eyes that relay information to your body's clocks about daylight are especially sensitive to blue light, but also respond to bright light. Blue light is abundant during daylight hours and has an arousing effect. This effect can have some intriguing applications. A small, randomized, controlled study has shown that wearing blue light-blocking glasses can significantly reduce symptoms of mania in patients being treated for bipolar disorder.[20]

TURNING UP MELATONIN AT NIGHT

- Dimming the lights so that ambient light is no brighter than 323–538 foot-candles and blocking out blue light a few hours before bedtime will encourage your body to start producing more melatonin in the evening.[21] Blue light–blocking glasses are becoming increasingly available and can help to maintain melatonin production during travel and shift work.

PRODUCING MELATONIN DURING THE NIGHT

- Bright daylight exposure during the day improves your melatonin levels at night.
- Bright or blue light at night can interrupt melatonin production, but you can reduce this effect by exposing yourself to bright daylight during the day. The less daylight you are exposed to during the day, the more susceptible you are to bright or blue light interfering with melatonin production during the night.[22] [23]

TURNING DOWN MELATONIN IN THE MORNING

- Morning light exposure curbs melatonin production in the morning. Dawn light exposure is especially good for your mind. Exposing your eyes to dawn light upon waking—or even a simulation of it—can improve cognitive performance compared to exposing your eyes to blue light.[24]

USING LIGHT TO BEAT JET LAG OR ADJUST TO SHIFT WORK

You can use light to reset the time on your clock if you are travelling.[25] Light can change the time on the body's clock, but whether it pushes the time forward or backward might depend on the time at which your body is at its coolest. Your body's temperature oscillates to a circadian rhythm. You cool down during the night and reach a minimum temperature in the early hours of the morning. In one study, light exposure six to eight hours before this point effectively pushed the clock back, and exposure between two and four hours after this point brought the clock forward.[26] These effects can vary considerably depending on light intensity, duration, and individual circumstances. When using light to change the time on your clock, the duration of light may play a more significant role than its brightness.[27] Full-spectrum white light works well, although blue light may have a more potent effect.[28]

- Get daylight exposure at every opportunity.
- Aim for at least three exposures—morning, noon, and afternoon.
- Turn your lights down from sundown onward. The best light is candlelight!
- Wear blue-blocking glasses every evening, and use light-filtering apps on electronic devices.
- Sleep in darkness with an eye mask and earplugs if your room is not completely dark or soundproof.

Food

WHAT TO EAT TO BOOST MELATONIN

Your body uses the amino acid tryptophan and vitamin B6 to make melatonin so you must not be deficient in either of these. Eggs and bananas are examples of foods that contain tryptophan and vitamin B6. A Japanese study showed the following combination to have a powerful effect on melatonin secretion at night.[29]

a. Very dim, low temperature lights in the evening/night before
b. A breakfast containing the amino acid tryptophan and vitamin B6
c. Exposure to more than thirty minutes of sunlight after breakfast

Eating a tryptophan-rich breakfast *without* bright-light exposure during the day did *not* have as large an effect on melatonin production as did the combination of a tryptophan-rich breakfast with bright-light exposure in another study.[30]

Some foods may contain "ready-made" melatonin that you may consider having as part of your supper, although the content can vary:

- Sprouted legumes[31]
- Tart Montmorency cherries[32]
- Milk from cows (with good circadian rhythms) milked at night[33]

In summary:

- *Make sure you get enough tryptophan and vitamin B6 (e.g., from bananas, eggs).*

- *Follow your breakfast with at least thirty minutes of daylight exposure.*
- *Include tart Montmorency cherries or sprouted legumes or nocturnally produced milk in your supper in the evening.*

WHEN TO EAT

Your body is optimized to eat during daytime hours. The rate at which your stomach empties and your intestines move things along is at its highest in the morning. Your gallbladder secretes bile into your intestines to help with fat digestion and mouse studies suggest this process is primed for daytime. Your gut bacteria also operate to a circadian rhythm and prefer not to have to digest food in the middle of the night.[34] [35] [36] [37] A study on shift workers at the British Antarctic Survey Station at Halley Bay in Antarctica showed that the levels of glucose and insulin circulating in the blood after eating a meal during a night shift are higher than when having the same meal during daytime hours, suggesting a form of insulin resistance develops during the night.[38] Melatonin prevents insulin from regulating your blood sugar.[39] So, dining late in the evening, in a dimly lit room, may be unwise. As you dim your lights, stop eating.

The clocks around your body respond to your "feeding times."[40] They work best on a fasting–feeding cycle where you feast during daylight and fast during darkness. Mouse studies suggest the bigger your feast, the longer should be your fast beforehand. The first meal of the day after a long overnight fast sets the time for morning.[41] Eating a heavy meal late in the evening shortens this overnight starvation interval and disturbs the rhythm.[42] You may be able to change the time on your peripheral clocks by eating in a different pattern for just a week.[43] Your sympathetic and parasympathetic systems operate to circadian rhythms, and starting the day with breakfast just five hours earlier than usual can make your parasympathetic tone shift its rhythm.[44] Eating later in the day can disrupt the natural rhythm of cortisol.[45]

What happens to the food you eat once it enters your body seems to depend on when you eat it. If you indulge in junk food, you are less likely to suffer its consequences if you avoid eating it in the evening or at night. Should you decide to have an unhealthy meal, it is far better to do so for lunch than at one in the morning while staying up to meet a deadline.

An experiment carried out on nocturnal mice (who sleep during the day and are awake during the night) revealed an extraordinary finding. The mice were given an unhealthy diet containing a set amount of total calories and split into two groups. One group was fed this diet during a fixed interval (an eight-hour time window) during the mice's natural waking hours. The second group was allowed to eat at any time during the entire twenty-four hours, during both natural sleeping and waking hours. The time-restricted group were bizarrely protected from becoming obese, which their free-feeding counterparts became. At the end of the study, the free-feeding mice weighed 23 percent more and had 70 percent more fat deposit than their time-restricted friends, despite eating identical diets with identical calories.[46]

- *Have your first substantial meal of the day in the morning.*
- *Precede it by a long fast.*
- *Eat your last meal as early as you can, in the evening.*
- *Stop eating as you dim the lights.*

■ *Front-load your day with more food in the first half of your day and as little as possible in the evening.*[47]

How to Use the Feeding–Fasting Cycle to Resist Jet Lag

It may be possible to use the feeding–fasting cycle to combat jet lag. A few days before you depart, start having your biggest meal of the day at the time when it is breakfast at your destination. Precede this meal with an eight- to ten-hour slot of not eating.

☺ EXERCISE

Exercise can increase cortisol, and it can also increase your sleep debt. Manipulating both of these effects may help you regulate your body clock.

When you exercise makes a difference. A few hours before you would usually sleep, your body starts wanting to sleep because you have clocked up a big "sleep debt." Your body clock overrides your urge to fall asleep and keeps you awake for another few hours. If you exercise *during* this window, you might push your clock back. Exercising in the morning versus the evening or not exercising at all have very different effects on melatonin, sleep pattern, body temperature, and your parasympathetic signals.[48]

■ If you don't exercise, your evening melatonin production may be reduced.
■ Your body temperature cools down during the night as you sleep. This cooling is impeded if you exercise in the evening.

- Your parasympathetic tone is increased during the night if you exercise in the morning.
- Your sympathetic tone is enhanced during the night if you exercise in the evening.
- The rapid eye movement phase of sleep is important for forming new memories. This phase may be *reduced by as much as 10 percent* from exercising in the evening.

In summary:

- *Exercise in the morning.*
- *If you must exercise in the evening, avoid exercising for at least three hours before bedtime.*
- *If you need to stay awake or push your body clock back, exercise in the evening within three hours of your usual bedtime.*

@ HEAT

Our bodies get coldest during the night and become warmer as we wake up and start to move around. We are more likely to fall asleep as our core body temperature drops . . . but there is a catch! We all live in two bodies. One is our outer "shell" of arms and legs and skin. The other is our inner "core," the cavity inside us that houses our organs. It is the "core" that gets colder as we drift off to sleep. This may be why cooling down after a hot bath before bedtime can make us feel sleepy. All that heat has to go somewhere, so it is "shifted" to the outer "shell," which heats up. The "shell" then dissipates the heat into the air around us. This provides the opportunity for a clever adjustment: temporarily heating up our "shell" confuses the brain and can make you fall asleep faster. This heating process must not interfere with the body's "core" cooling down. Your "core" needs to become cooler in order to sleep. So a comforter folded down to cover just your lower legs and feet will help you drift off faster. Foot warming has been shown to bring on sleep.[49]

- *Keep your bedroom cool.*
- *Take a hot bath before bed.*
- *Cover only your lower legs and feet with a comforter.*
- *Or, alternatively, wear warm socks!*

⊚ ALERTNESS

Melatonin opposes the stress signal, so anything that makes you alert or that excites you—whether social company, television, or loud noise—may act against melatonin. Wear earplugs during the night if your room is not soundproof.

Consuming caffeine in the evening will keep you awake for longer than caffeine consumed during the day. In one experiment, drinking the equivalent of a double espresso three hours before bedtime shifted melatonin rhythm by 40 minutes.[50] Excessive caffeine can cause harm, so it may be safer to take caffeine in the form of a cup or two of tea or coffee, instead of the tablet form. Do not take too much. According to the Mayo Clinic, a maximum of 400mg of caffeine per day is thought to be safe for adult consumption, which approximates to four cups of brewed coffee.[51] *Too much caffeine is dangerous for your health.*

- *Avoid sources of loud noise and excitement from the evening onward.*

- *Stop taking caffeine-containing drinks more than three hours before bedtime.*
- *Wear earplugs while you sleep, if necessary.*

⊚ PUSHING YOUR CLOCK BACK TO STAY UP LATER

- Expose your eyes to bright light late in the evening.
- Exercise. In order to stay awake longer, a good time to exercise is within the few hours before you usually fall asleep.
- Keep warm. Make sure your whole body stays warm, though, not just your feet!
- Drink a cup of coffee in the evening.

Shift Work

- For day–night reversal, wear blue light–blocking glasses in the morning and try to keep wearing them throughout the day. Take them off in the evening.
- Have your largest meal of the day after an eight- to ten-hour fast, when you want it to be "morning."

⊚ PUSHING YOUR CLOCK FORWARD TO FALL ASLEEP EARLIER

- Make your lights dim hours before you want to fall asleep and block blue light.
- Wear blue light–blocking glasses if you are watching movies on a plane or if you are working all night on your computer.
- Wake up an hour or two earlier than usual and expose yourself to daylight or to bright light if it is dark.

- Eat breakfast earlier
- Exercise well to make yourself tired and accumulate "sleep debt" but avoid the three-hour slot before you want to fall asleep.
- While you try to fall asleep, start to cool down your body with cool circulating air, but warm up small parts of your shell by wearing warm socks or folding the comforter to cover your lower legs. Take a hot bath.
- Avoid caffeine.

AVOIDING THE POST-LUNCH DIP

A *post-lunch dip* is the term for feeling sluggish after lunch. It can affect mental performance and cognitive flexibility. Various studies have explored ways to get rid of the post-lunch dip, which happens even to people who have slept well the night before and are generally well rested. Studies have identified the following as helpful:

Blue Light or Bright Daylight

Half an hour of bright daylight exposure after lunch can help to reduce a post-lunch dip.[52] In one experiment, people exposed to artificial light "enriched" with blue light in the "post-lunch period" were more alert and their mental performance was enhanced.[53]

A Short Nap after Lunch

If possible, consider a short power nap.[54] A forty-five-minute nap may shift you toward a higher parasympathetic tone than a fifteen-minute nap, and you may perform slightly better on a cognitive test after a fifteen-minute nap.[55] Although a nap will help you work better afterward, it might take a little while to return to a high level of concentration if you are doing something that needs analytical skills.[56] A nap also improves your mood.[57]

A Smaller Meal

Eating lunch at all makes the dip worse.[58] A large lunch is worse for the post-lunch dip than a smaller lunch.[59]

- Get half an hour of daylight.
- Avoid eating too large a lunch.
- Take a short nap, if you are able to.

◎ AN OPTIMAL DAILY ROUTINE

MORNING
- Wake up and expose your eyes to morning light.
- Eat a substantial breakfast at the same time, every day.
- After breakfast, get some daylight for at least thirty minutes.

DAYTIME
- Try to get at least half an hour of daylight at three separate times of the day: after breakfast, at lunchtime, and in the afternoon.
- Exercise in the morning. If you can't exercise in the morning, exercise in the afternoon.

EVENING

- Eat your evening meal as early as possible. Keep it small.
- Wear blue-blocking glasses.
- Use programs such as f.lux to reduce blue-light emissions from electronic devices.
- Dim your lights.
- Stop eating as you dim your lights.
- Avoid intense exercise.
- Avoid noisy television.
- Avoid excitement.
- Avoid scrolling through social media or watching things that make you react with anger or negativity.
- Avoid caffeinated drinks and alcohol.

As YOU HEAD TO BED

- Switch off electronic devices.
- Read a real book.
- Take a hot bath.
- Keep your room cool and keep your feet warm.

CHAPTER 6

Extinguishing Inflammation

YOUR BODY HAS a battalion of foot soldiers marching over every region, scanning for threats. They form part of your immune system. If an enemy has ventured in, the foot soldiers light a fire to incapacitate it. The fire generates heat and a red glow. It also produces pain and swelling. In the first century AD, Aulus Cornelius Celsus, an encyclopedist in ancient Rome, wrote that fire within the body had four cardinal signs: *calor* (warmth), *dolor* (pain), *tumor* (swelling), and *rubor* (redness). Whether the fire is as delicate as a candle's flame or as vicious as a raging inferno, it is there to defend you from enemy agents. That fire is what we call *inflammation.*

Like foot soldiers on the lookout for an invading enemy, your immune system is always scanning your body for unwelcome guests. The fire that it lights may be large or small, widespread or localized. If your immune system decides the enemy is too strong, fire rages across your body. You feel this as a fever. If your enemy has been contained, inflammation stays localized. A cut in your finger may become red, swollen, and painful but you otherwise feel well. Sometimes, a subtler form of systemic inflammation can act more insidiously. You may not have a fever, but you don't feel quite right.

Certain lifestyle factors can encourage inflammation, and many of us may be in a state of chronic, low-grade inflammation without being aware of it.

Many of us may be in a state of chronic, low-grade inflammation.

Stress has an intricate relationship with inflammation. If we split a stress response into the autonomic response and the HPA-axis response, we find an interesting interplay. Sympathetic activation seems to encourage inflammation and parasympathetic activity has a calming effect, although in reality their interplay is rather more complex.[1] [2] While sympathetic activity may be inflammatory, the stress hormone cortisol reduces inflammation.

These interactions converge into the tendency toward rising inflammation in the setting of chronic stress and a decline in inflammation with release from stress. In a study involving more than six hundred middle-aged individuals, going through money worries over the previous year correlated with signs of inflammation.[3] Cancer care-givers have significantly higher levels of the marker of inflammation C-Reactive Protein (CRP), compared to people who are not under a similar degree of stress.[4] A small study looking at sixteen healthy young female medical students found an inverse correlation between HRV measurements and levels of inflammatory markers such as tumor necrosis factor-alpha (TNF-α).[5]

INFLAMMATION AND THE BRAIN

Inflammation, stress, and depression incite each other into a self-perpetuating cyclone. Inflammation can trigger the stress cascade.[6] [7]

Inflammation can trigger the stress cascade.

Stress may also increase inflammation.[8] [9] [10] [11] [12] [13] [14] The presence of inflammation or the body's natural inclination toward it can determine whether or not a stressful experience will eventually lead to depression. We know, for example, that the body's tendency to produce the inflammatory agents IL-1β and IL-6 correlates with the severity of depression brought on by interpersonal stress.[15]

Making healthy people inflamed makes them depressed.[16] [17] Some cases of major depression are treated more successfully with *anti-inflammatories* rather than with antidepressants.[18] Inflammation is linked to a decline in mental performance and some inflammatory agents seem to specifically target the rational brain, interfering with brain circuits involved in cognition and memory. [19] [20] [21] [22] [23] [24] [25] People suffering from some inflammatory illnesses can experience a "foggy brain" or "brain fog," which subsides when the inflammation improves.[26] [27] [28] [29] [30]

⊙ STRESS AND INFLAMMATION

The vicious cycle of inflammation and stress loses its momentum as soon as you reduce the load of either inflammation or stress. Stress is something you cannot control, but if you do all you can to calm inflammation and avoid enkindling it in the first place, you will provide a strong counterbalance as stress pushes you toward inflammation.

We have identified several routes through which inflammation can strike. Blocking these routes will help you buffer the inflammatory effects of chronic stress.

These routes relate to:

- The integrity of your intestinal walls
- Your microbiota

- General nutrition
- Lifestyle

The Integrity of Your Intestinal Walls

Your intestines provide the ultimate barrier between the outside world and you. This barrier is made of two parts: the structural wall of your intestines and the layer of mucus that lines it. The intestinal wall is like a brick wall composed of lots of bricklike cells that are cemented together by structures known as **tight junctions**.[31] A thick layer of mucus lines the walls of your intestines like wallpaper, providing an extra barrier between its contents and the rest of your body.

The intestinal barrier is a critical checkpoint for your immune system and large numbers of its soldiers gather here, making sure disease-causing bacteria, viruses, and fungi don't sneak into your body. Anything that creates a hole in either the mucus layer or the intestinal wall, or both, will allow unwelcome debris to slip into your body and cause inflammation. A gut with a compromised wall is sometimes called a "leaky gut."

Lipopolysaccharide endotoxin (LPS), sometimes referred to as endotoxin, forms part of the outer structure of "bad" bacteria. It is recognized as an "enemy" by your immune system and it triggers systemic inflammation if it crosses into your blood through cracks in the gut wall. The degree of inflammation it triggers depends on how much is entering into your body. A relentless trickle of a small amount of endotoxin may not cause obvious, blazing inflammation but it can evoke symptoms of depression, lower mood, and increase anxiety levels.[32][33] Chronic fatigue syndrome is associated with stress and CFS sufferers may have higher levels of endotoxin circulating in their blood than non-sufferers.[34]

FACTORS THAT MAY CAUSE DAMAGE TO
THE INTESTINAL BARRIER

The following factors can either cause or encourage a leaky gut:

- Psychological or physical stress
- A poor diet
- Medications including NSAIDs and PPIs

PSYCHOLOGICAL AND PHYSICAL STRESS

Stress disrupts your intestinal barrier. Something as benign as speaking in public can result in a leaky gut.[35] Intense exercise makes the intestinal walls more permeable; endurance exercise such as marathon running is well known for causing unwanted gastrointestinal symptoms. Heat exposure also increases intestinal permeability and intense exercise in the heat poses a two-pronged attack on the intestinal barrier.

Stress is likely to make a gut leaky through multiple avenues, many of which remain unknown. One interesting theory regarding stress and intestinal permeability involves the vagus nerve. The vagus nerve is one of the principal conduits for parasympathetic input to your body from your brain. The activity of the vagus nerve, or "vagal tone," can be measured to estimate how stressed you are.

One of the many horrific consequences of suffering from a severe burn injury is increased intestinal permeability. If you look at the intestinal wall under a microscope after a burn injury, you will find large gaps between the cells lining the wall. One mouse experiment has shown that stimulating the vagus nerve within sixty minutes of a burn injury preserves the integrity of the tight junctions and prevents the breakdown of the intestinal wall. No gaps are seen under the microscope. Stimulating the vagus nerve ninety minutes after a burn injury preserves many, but not all of the tight junctions and you can see small spaces between the cells.[36] If vagal tone is increased quickly after a stressful experience, this seems to prevent cracks from forming in the intestinal wall.

Regular exercise can help to reduce your overall stress reactivity, but as we've seen, *intense exercise* is stressful. This isn't a problem if you exercise for short bouts, but if you exercise intensely for too long you are keeping your vagal tone lowered for longer. Recovery after intense exercise also takes time. This has been proposed as a possible theory behind why *intense* exercise causes intestinal permeability and gives rise to chronic inflammation, which can do more harm than good.[37] Various forms of physical stress, including heat stress and combat training, are believed to trigger gut leakiness by the same mechanisms.[38] Some scientists are wondering if this role of the vagus nerve might explain why exercising too intensely can cut a life short.[39]

A POOR DIET

General observations and laboratory experiments have identified some dietary factors that may damage the intestinal barrier.

▪ Alcohol is best minimized or avoided altogether in the setting of stress. An excess of alcohol may encourage brain inflammation, or **neuroinflammation**.[40] Alcohol increases intestinal permeability, perhaps through its effect on gut bacteria. Gut bacteria play a central role in the genesis of alcohol-related liver disease.[41] Intriguingly, feeding rats about 5g per pound of body weight of Quaker Oats twice a day while they were given 4g per pound of body weight of ethanol per day *prevented them from developing a leaky gut.*[42] The reason for this may rest with the beta-glucan, the soluble fiber found in oats.[43] Soluble fiber strengthens the integrity of the mucus layer lining the intestinal wall. Alcohol-induced gut leakiness may be made far worse by not producing enough melatonin at night. Binge-drinking and staying up late is a terrible combination for your intestinal barrier and for inflammation.[44]

- When mice were fed a diet of 25 percent lean red meat or casein, the principal protein in cheese, their colonic mucus layer became thinner compared to its thickness when they were fed 15 percent casein, suggesting that high levels of these proteins may disrupt the intestinal barrier. This thinning was prevented by replacing the starch that formed 48 percent of the diet with resistant starch.[45] This finding suggests that if protein forms a large proportion of your diet, increasing your intake of resistant starch may help protect your intestinal barrier. Resistant starch is starch that gets digested more slowly than plain starch. Green bananas contain more resistant starch than ripe, yellow bananas. Cooked white rice that is cooled for twenty-four hours at 39.2 °F then reheated changes its resistant starch content from 0.64g/100g to 1.65g/100g compared to freshly cooked white rice.[46]

- Animal studies suggest a diet that is high in fat may disrupt the intestinal barrier in part through the dynamics of bile acid secretion. In one experiment, a diet of 40 percent fat, where the fat consisted of a mixture of lard and soybean oil, increased intestinal permeability when compared to a control diet that contained 7 percent fat from soybean oil. The increase in permeability correlated with bile acid secretion.[47] Bile acid secretion also affects gut bacteria. It may be possible to offset these effects, at least in part, by including dietary fiber and starch within the meal. These bind onto the bile acids so they are not floating "free" within the intestines. There is a theory that bound bile acids may cause less harm to the gut than "free" floating bile acids.[48]

- If you eat dairy cream, you might be able to make a difference in intestinal permeability by choosing cream from pasture-raised cows, rather than a more standard variety of cream. We aren't quite sure why, but a mouse experiment has shown the former keeps the intestinal barrier more stable.[49]

- You should avoid artificial sweeteners and emulsifiers as they can damage your intestinal barrier and make your gut leaky.[50] [51] [52] [53] [54] Diet soda drinks are a prime culprit and should be avoided entirely.

- There is a food group known as FODMAPs that may increase intestinal permeability in the context of irritable bowel syndrome (IBS). More on this later.

MEDICATIONS INCLUDING NSAIDS AND PPIS

Non-steroidal anti-inflammatory drugs (NSAIDs) such as aspirin and indomethacin increase intestinal permeability. This effect is made a great deal worse with the coexistence of psychological stress. If you are going through stress, be particularly conscientious about looking after your gut health if you must take painkillers.[55] Part of this effect is mediated by changes in gut bacteria and this will be described in more detail shortly. The worst possible combination for intestinal permeability is to have a stressful job, stay up late, or work night shifts and then binge drink on weekends with some aspirin the next day to calm a hangover.

Proton-pump inhibitors (PPIs) that many people use for symptoms of acid reflux may also increase gut leakiness.

Preventing a Leaky Gut

The following factors may help keep the gut lining sealed:

- Dietary soluble fiber
- Glutamine
- Reducing the duration of a stress response
- Plant substrates
- Vitamin D

DIETARY SOLUBLE FIBER

Soluble fiber strengthens the integrity of the mucus layer that lines the intestinal wall. This may explain why mice don't seem to develop a leaky gut from drinking alcohol if they eat oats twice a day.[56] [57] Beta-glucan is the soluble fiber found in oats. Other examples of soluble fiber include psyllium husks, the husk of the seeds of the plant *Plantago ovata* whose use is widespread in India, pectin, found in apples, and alginate, found in brown algae and used widely in Japan.

GLUTAMINE

The cells lining your gut feed on the amino acid called glutamine. They pick their food from what travels down from your stomach, so the food you are swallowing needs to contain this.[58] Milk protein contains about twice as much glutamine as beef protein and can be a good source of it.[59]

REDUCING THE DURATION OF A STRESS RESPONSE

Earlier, I described a theory that is based on the ability of the vagus nerve to encourage the integrity of the tight junctions between the cells of the intestinal wall. Stimulating the vagus nerve within sixty minutes rather than ninety minutes of a burn injury prevents cracks from appearing in the intestinal wall in mice. Although this remains an unproven theory, it suggests that a swift return to a relaxed state after a stress response is likely to help preserve cohesion of the intestinal wall. Please refer to chapter 5 for a reminder of the strategies that can help you do this.

PLANT SUBSTRATES

- You need zinc to repair cracks in the gut's wall. It works along both the large and the small intestines.[60] [61] Nuts, vegetables, and meat all contain zinc.
- Increasing your intake of a variety of plants in the form of fruits, vegetables, and whole grains will provide you with resistant starch in addition to soluble and insoluble fiber, all of which can encourage the integrity of your intestinal lining.[62]
- The spice turmeric may calm inflammation within the intestines.[63] There are currently several clinical trials in place looking at treating Crohn's disease and ulcerative colitis, which are both inflammatory diseases of the bowel, with turmeric.[64]

VITAMIN D

A deficiency in vitamin D is linked to intestinal permeability, and vitamin D supplements have improved gut leakiness in the setting of Crohn's disease.[65] [66] [67]

Your Microbiota

Over 10 trillion bacteria as well as viruses (known as phages) and fungi live within your digestive tract.[68] [69] Collectively, this ocean of creatures is referred to as the *microbiota*. In the context of stress and inflammation, your microbiota has three key roles.

First, it prevents microbes that make it into your digestive tract from entering into your body and causing inflammation (and disease). It does this in two ways.

- It nurtures and maintains the intestinal walls and the mucus layer.
- Your resident germs try to overpower the foreign germs entering into your gut.

Second, your microbiota shifts the balance of your immune cells so they are less likely to trigger unnecessary inflammation.

Third, there is some evidence that the bacteria in your gut may communicate with your brain and modify your stress response.

BUILDING A HEALTHY MICROBIOME

Research into the microbiota is still young and there are many gaps in our knowledge. We do know, however, that we should be aiming toward *diversity*. A diverse microbiota is resilient and can adapt to change. A diverse ecosystem supports organisms with a variety of talents and traits, which can together put up a more successful fight when an invading bacterium enters your intestines and threatens to cause disease.

Diversity breeds diversity. If you want to nurture a diverse range of microbes, you need to provide your microbiota with a diverse range of foods. Your gut bacteria feed on **microbiota-accessible carbohydrates**, or **MACs**, which are carbohydrates that are difficult to digest. Dietary fiber is a source of MACs and eating a diverse range of fiber sources can help to nurture a healthy microbiota.

Basing your diet on only one MAC is unwise because different bacteria thrive on different substrates, so if you only eat one thing you will nurture a narrow band of bacteria excessively while neglecting others, and this can have negative consequences.

The microbiota responds very quickly to a change in diet, and bacterial populations can change in as little as twenty-four hours.[70] Unprocessed, whole, fiber-rich plant foods contain a complex variety of MACs, whereas processing food removes multifariousness to produce bland homogeneity. Processed-food intake is not good for a healthy microbiota. At every meal, you should aim to provide your gut bacteria with a miscellaneous array of unprocessed, whole, plant-derived foods.

If you usually eat processed, polished white rice, you are waylaying the opportunity to strengthen your intestinal barrier. Feeding animals rice bran increases levels of *Lactobacillus rhamnosus GG* and improves the cohesion of the intestinal barrier. This prevents animals from suffering from diarrhea when they are exposed to the diarrhea-causing virus rotavirus, preventing inflammation and disease.[71] Giving healthy humans either brown rice or whole-grain barley or both for four weeks reduced levels of the inflammatory agent IL-6 in a randomized crossover trial.[72]

There is a receptor known as the aryl hydrocarbon receptor, or AHR, which plays a critical role in preventing inflammation by protecting the intestinal barrier and modulating immune cells. The receptor is activated by metabolites produced by your gut bacteria, highlighting the need for a healthy microbiota. It can, however, also be stimulated by several plant-derived ingredients including cruciferous vegetables, flavonoids, and polyphenols.[73] Vegetables and fruits are good sources of minerals such as magnesium and zinc, which are both good for healthy intestines. Putting mice on a low-magnesium diet for six weeks changes their microbiota and they become depressed.[74] A diet that is rich in colorful vegetables and fruit shows an inverse correlation with inflammation.[75] Both a Mediterranean diet and a Paleolithic diet, which include a variety of plants, have been associated with lower levels of inflammation.

- Feed your gut bacteria *a wide selection of plant-derived sources of fiber* every day.
    - Include a small representative portion from different plant groups, such as a small portion of seeds, nuts, legumes, cruciferous vegetables, tubers, root vegetables, fruits, spices, and grains.
- Avoid processed food.

The benefits of fruits and vegetables reach far beyond their effect on the gut microbiota. Plants contain a wealth of phytonutrients, which are associated with improved brain health. One good strategy is to adopt a "rainbow plate" with fruits, vegetables, herbs, and spices that reach across the color spectrum, because different plant pigments can bring unique advantages to the table.[76 77 78 79 80 81]

THE NEED FOR PROBIOTICS

Your intestinal microbiota is fighting a constant battle with insurgents entering through your food and drink. Troop numbers decide victory in a battle, so regularly adding "good" gut bacteria to your armory will help your microbiota overcome insurgency.

The different bacterial species in your intestines have their own names. You can identify a bacterium's tribe from its name. Tribes that produce lactic acid have the name Lactobacillus.

Lifestyle factors can diminish bacterial populations. Antibiotic use is one example. *Lactobacillus* strains are a famous casualty of the widespread use of penicillin-related antibiotics, and a new generation of antibiotics is being designed to take the shape of a dual-compartment pill that incorporates *Lactobacillus* bacteria.[82]

Stress is another lifestyle factor that can affect gut bacterial populations. Stress and *Lactobacillus* strains exist in perpetual opposition. If one rises, the other falls. If some *Lactobacillus* populations go up in number, stress goes down, but if stress goes up, their numbers go down.[83][84]

- Some strains of *Lactobacillus* appear to actively bring down the stress signal. Injecting the intestines of mice with *Lactobacillus johnsonii La1* reduced their blood pressure and raised parasympathetic tone.[85] *Lactobacillus helveticus* also has this effect.

- In a randomized, double-blind, placebo-controlled study on 171 volunteers, taking yogurt that contains *Lactobacillus plantarum* for two months reduced markers of stress.[86]

- Giving chronic fatigue syndrome sufferers *Lactobacillus casei* daily in a fermented milk drink for two months raised both *Bifidobacterium* and *Lactobacillus* levels in the colon and significantly reduced anxiety in a small, randomized, double-blind, placebo-controlled study.[87]

- Taking a probiotic that contained *Lactobacillus helveticus* and *Bifidobacterium longum* for a month reduced symptoms of depression, anxiety, and anger and lowered cortisol in healthy adults.[88]
- The stress of taking exams reduces the population of *Lactobacillus* in college students.[89] *Bifidobacterium* numbers are also reduced during stress and some types of *Bifidobacterium* might help with anxiety.[90]
- A fermented milk drink (containing *Lactobacillus casei*) improved the mood of 132 healthy adults after three weeks, particularly if they were feeling depressed at the start.[91]
- A group of healthy women were given a fermented milk product made with yogurt starter cultures that contained *Bifidobacterium animalis* subsp. *lactis, Streptococcus thermophiles, Lactobacillus bulgaricus,* and *Lactococcus lactis* subsp. *lactis* daily for four weeks. At the end of the four-week period, the brains of those taking the fermented milk product reacted differently to painful and emotional stimuli, compared to the brains of matched controls.[92]
- If you colonize the intestines of mice with *Lactobacillus rhamnosus,* there is an improvement in stress-induced anxiety and depression.[93]

Over a century ago, the Nobel Prize–winning immunologist Ilya Ilyich (Élie) Metchnikoff, one of the founding fathers of inflammation research, proposed that eating yogurt every day is the secret to a long life after noticing an astonishing number of frugal centenarians living in the Balkan states who ate yogurt regularly. Although we need more studies, it would appear that current research findings support this view.[94] [95] People with lactose intolerance often find they are able to eat yogurt without any problems, since the lactose sugar is broken down by the active cultures. If you are allergic to milk, sauerkraut (European fermented cabbage), natto (Japanese fermented soy), and kimchi (Korean fermented cabbage) are other options.

■ Unless you are allergic to dairy products, take plain, unheated, probiotic yogurt every day. Aim for approximately 300g per day.[96] Exchanging yogurt for your usual dessert at the end of every meal may be an effective strategy.

If you have noticed you become ill more easily when you are chronically stressed, your gut bacteria may be partly to blame. When you are stressed, noradrenaline circulates around your gut. It increases the virility of several pathogenic bacteria. For instance, it makes the gastroenteritis-causing bacteria E. coli multiply rapidly.

Irritable Bowel Syndrome

A hundred million nerves congregate along your gut.[97] Just as the brain in your head can respond to chronic stress with depression, the "gut brain" can respond to chronic stress with irritable bowel syndrome.[98] [99] There is a strong link between IBS and exposure to stress and stressful occupations.[100] [101] Its prevalence varies across countries and cultures and has been reported to be as high as 48 percent in female African medical students and 41 percent in male Korean medical students about to begin their residency.[102] [103] IBS may relate to an imbalance in the microbiota that has been caused by stress. One study has shown how going through combat training changed the microbiota of healthy male soldiers and the change correlated with an increase in intestinal permeability and symptoms of IBS.[104]

IBS remains poorly understood and is a "diagnosis of exclusion." It is possible that several different disease processes may be taking place under the common banner of IBS and these processes have yet to be identified. IBS can manifest in a kaleidoscope of symptoms, from bloating, heartburn, and acid reflux, to excessive gas and constipation, diarrhea, or all three. The symptoms can be mild or severe, fleeting or persistent.

It is wise not to ignore IBS, for two reasons. First, IBS seems to raise markers of inflammation in some cases.[105] Second, having IBS increases the risk of depression and anxiety, implying that treating it may reduce this risk.[106]

There is a theory that disordered gut bacteria may ferment a specific kind of carbohydrate in the diet (known by the acronym FODMAPs) excessively and the products of fermentation may cause undue distress, leading to the typical symptoms of IBS.[107] IBS may be controlled by reducing FODMAPs, as demonstrated in at least two randomized, controlled trials.[108] [109] A low-FODMAP diet also reduces levels of inflammatory markers.[110] There have been case reports of endurance athletes who suffer from gastrointestinal symptoms deriving benefits either from reducing FODMAPs in their diet, or by avoiding FODMAPs during intense stressful training.[111] [112]

Intestinal inflammation and an imbalance of immune cells may also contribute to IBS. In addition to FODMAPs, wheat contains proteins such as gluten and amylase-trypsin inhibitors that can elicit inflammation and worsen intestinal permeability.[113] Many IBS sufferers are intolerant to wheat and may suffer from symptoms of discomfort when eating some other types of grain, such as rye and barley.[114] [115] [116] [117] Fibromyalgia is a musculoskeletal illness that has been linked to stress. It can coexist with IBS.

In one study, treating fibromyalgia patients with a low-FODMAP diet reduced their muscular pain.[118] [119]

Low-carbohydrate diets are growing in popularity, with their proponents often anecdotally reporting improved digestion and general well-being. It is possible that these benefits may stem, in part, from the inadvertent elimination of wheat and FODMAPs as a consequence of restricting overall carbohydrate intake in the setting of a preexisting IBS landscape.[120]

If you have any issues with your digestion, whether they are trivial, acute, or long-standing, I would suggest a visit to your physician to exclude an undiagnosed illness and rule out IBS. If you are diagnosed with IBS, you may find it helpful to adopt a low-FODMAP approach to your diet, under the guidance of your physician, particularly during periods of intense stress.

Although insoluble fiber is not always helpful in IBS, soluble fiber can help its symptoms. Psyllium husks, the husks of the *Plantago ovata* seed, contain soluble fiber. In one study, 0.35 ounce of psyllium husks taken twice a day significantly improved symptoms of IBS.[121] Since IBS may result from disarranged gut bacteria, replenishing your microbiota with probiotics may also help.[122] In addition to eating plain, unheated, probiotic yogurt every day, you may wish to consider taking a short course of probiotics after discussing this with your physician. VSL#3 is a probiotic supplement that has been shown to be effective at improving symptoms of IBS in children, in a randomized, controlled trial.[123] In summary, three interventions to carefully explore with your physician in the context of IBS are:

- Reducing FODMAPs and avoiding wheat- and gluten-containing products
- Psyllium husks
- Probiotics

Please note: It is essential to thoroughly discuss *all* aspects of this and any other dietary approaches with your physician *before* you make any changes to your diet. IBS is a clinical diagnosis and must not be self-diagnosed.

⊚ NUTRITIONAL FACTORS

The simple act of eating can trigger inflammation, perhaps in part through effects on gut bacteria and intestinal permeability. Scientists from the University of South Carolina have rigorously put together results from 6,500 published papers to create a "Dietary Inflammatory Index" that lists the tendency of different foods to cause inflammation.[124] It is possible to closely predict how inflamed we are on the basis of our dietary habits, using this Dietary Inflammatory Index.[125] The overall inflammatory effect scores of some fat parameters are listed as follows (the higher the number, the more inflammatory the food):

- Saturated fat = 0.373
- Trans fat = 0.229
- Total fat = 0.298

Some foods are cited as having an anti-inflammatory effect (the more negative the score, the more potent the anti-inflammatory effect):

- Turmeric = −0.785
- Dietary fiber = −0.663
- Flavones (found in celery, peppers, and thyme) = −0.616
- Isoflavones (found in unprocessed soy beans and other beans) = −0.593

- Beta-carotene (found in carrots and sweet potatoes) = – 0.584
- Green/black tea = –0.536

Saturated Fat and Trans Fat

Like necklaces, saturated fats can come in three different chain sizes: short, medium, and long. Long-chain saturated fats are different from their short- and medium-chain cousins.[126] There is evidence to suggest that one, in particular, may cause inflammation.[127] [128] [129] Palmitic acid or palmitate is found in animal fats and in palm oil. It constitutes about 43 percent of palm oil. Milk fat contains about 30 percent palmitic acid, and about 25 to 28 percent of lard is palmitic acid. Palmitic acid forms part of our bodies' structure and our bodies manufacture it. The difference between eating food that is high in palmitic acid and producing palmitic acid within the body is that eating it causes a transient *rise* in the level of palmitic acid circulating in the blood. This rise has the potential to cause inflammation.

After we have eaten a meal that is high in fat, the fats in our food become packaged as "postprandial triglyceride-rich lipo-proteins," or postprandial TRLs, which circulate in the blood. A large meal that is rich in palmitic acid will culminate in palmitate-heavy TRLs.[130] As your blood carries TRLs around the body, various vulnerable structures, such as the lining of blood vessels, will be exposed to the palmitate.

In a small study, six volunteers, aged between twenty-five and forty-five, were given either cow's milk cream, olive oil, or olive oil with omega-3 fish oil, on three separate occasions. A few hours after they ate, TRLs peaked in their blood and were analyzed. After eating cream, palmitic acid formed 36 percent of the fat in TRLs compared to around 12 percent after the olive oil options. Oleic acid, a monounsaturated fat, formed 23 percent of the fat in TRLs after the cream, versus 61 to 66 percent after both versions of olive oil.

The scientists conducting the study then tried to mimic what was likely to happen when these TRLs, circulating around the volunteers' bodies, were reaching their eyes. They took retinal cells, cultured them on a plate, and exposed them to different concentrations of the TRLs taken from the volunteers' blood.

The cream-related TRLs, which were higher in palmitic acid, caused far more oxidative stress and inflammation in the retinal cells than both of the olive oil–related TRLs, implying that any meal that delivers large amounts of palmitic acid into the blood has the potential to cause inflammation and damage retinal cells.

A near-identical study on fourteen healthy volunteers, using butter instead of cream, demonstrated a similar phenomenon taking place in the arteries that supply the heart.[131]

A meal that is high in long-chain saturated fat can also throw intestinal bacteria off balance within a very short time.[132]

The repeated heating or prolonged heating of cooking oils can produce *trans-fatty acids*, or *trans fats*, which are universally recognized as potent triggers of inflammation, heart disease, and cancer.[133] In one recent study, heating refined soybean oil, groundnut oil, olive oil, rapeseed oil, and clarified butter, as well as partially hydrogenated vegetable oil, to 356 °F increased the content of trans fats.[134]

RED MEAT

Components of one's diet may be relatively harmless in one setting and cause harm in another. Red meat may be one such example. A selection of epidemiological studies point toward a link between red meat and inflammation, which becomes stronger the more overweight you are, possibly because being overweight pushes you into an inflammatory state.[135] In an inflammatory paradigm, such as in a stressful lifestyle, small risks may become magnified.

> *In a stressful lifestyle, small risks may become magnified.*

- One potential culprit that contributes to the inflammatory effect of meat is the molecule Neu5Gc. It is uniquely present in red meat and uniquely absent in humans.[136] We don't naturally carry any Neu5Gc, so any that enters our body is recognized as an enemy agent. When you eat food containing it, it crosses into the body, and a small amount can become incorporated into your organs, including the liver. Since it is a foreign molecule, the *human* immune system attacks it, causing inflammation. Mice can be genetically engineered to not carry Neu5Gc. When these mice are fed Neu5Gc, their immune system gets activated and they suffer from widespread inflammation.[137] A recent study has shown people who suffer from autoimmune thyroid disease have significantly higher levels of antibodies to Neu5Gc than those who don't. There is a theory that proposes multiple sclerosis might be triggered from eating red meat because of this molecule.[138] [139] [140] There is more Neu5Gc in beef than there is in pork, and there is more in pork than there is in lamb. Although it is still mainly a theory, this theory is powerful because it explains why humans seem to be inflicted with a greater degree of inflammation from eating red meat than other carnivores. Other carnivores carry

Neu5Gc, so their immune system does not get excited if it enters their body.[141]

If you eat red meat, *eat small portions.*[142] The American Cancer Society recommends eating smaller portions of red meat or replacing red meat with fish, poultry, and beans wherever possible.[143]

@ LIFESTYLE FACTORS

High-Temperature Cooking

High-temperature cooking—such as frying—can generate more pro-inflammatory agents than low-temperature cooking.

- Cooking muscle meat at high temperatures creates compounds known as heterocyclic amines, or HCAs, which are carcinogens. The more cooked a piece of steak is, the more HCAs it contains.[144] Making burger patties with turmeric and rosemary as well as with olive oil, onion powder, and even hibiscus extract can reduce HCA formation.[145] [146] [147] The fat in meat can also become oxidized at high temperatures to form "oxidized lipids," which encourage inflammation. Cooking a steak at a low temperature for a long time produces fewer oxidized lipids than frying it in a pan.[148] Eating a juicy, fried steak results in more inflammation after the meal than eating one that has been cooked "sous-vide," a technique using lower temperatures (140 °F) over a longer period of time. The iron content in red meat may also contribute to inflammation but cushioning your meat in chlorophyll-rich green vegetables may help to reduce its negative effects.[149] [150]

- Cooking protein or fat in the presence of glucose (which may already exist in the food being cooked) induces a chemical reaction causing sugar molecules to react with and attach onto proteins and fat. The resulting product is known as an advanced glycation end product (AGE). AGEs might cause inflammation within our bodies when we eat them.[151] [152] Grains, legumes, breads, vegetables, fruits, and milk may be lower in AGEs than other types of food, unless they are prepared with added fats. Cooking meat that has been marinated in an acidic substance such as vinegar may reduce AGE production. Low-temperature cooking may be better than high-temperature cooking.[153] [154] Dry heat processing may accelerate AGE formation, and moist heat cooking is safer.

Exercise

Physical activity can reduce inflammation.[155] As you will discover later, being physically active is protective against insulin resistance. Avoid being sedentary.

Time-Restricted Eating

There is some evidence that spending a good portion of the day *not* eating may help to lower inflammatory markers.[156] This is known as **time-restricted feeding**. In one randomized study, those eating strictly within an eight-hour time window (at 1 P.M., 4 P.M., and 8 P.M.) experienced a fall in inflammatory markers after two months, compared to those who consumed the same number of calories spread across a wider time window (at 8 A.M., 1 P.M., and 8 P.M.).[157] Melatonin, the hormone we produce during overnight darkness, has an anti-inflammatory effect, and we turn off its production by eating a large meal late in the evening.[158] Eating within an eight-hour window may help to reduce inflammation.

Caloric Restriction

One of the kindest things you can do for your gut is to never overeat; neither in a single meal nor in an entire day, neither by volume nor by calories. Eating fewer calories than your body burns in a given day has an anti-inflammatory effect.[159] Caloric restriction may work in concert with your body clock, so if you look after your body clock you may reap even greater benefits.[160] You do not have to restrict calories, just make sure you don't eat more than you need. The Japanese have a custom of not eating any more when they "feel" 80 percent, *not 100 percent*, full. Known as *Hara Hachi Bu*, this cultural practice is thought to contribute to the extraordinary longevity seen in Japan.

Eating fewer calories than your body burns in a given day has an anti-inflammatory effect.

Visceral Fat

Visceral fat is the white fat nestled in your abdomen, in between your abdominal organs, rather than the fat that lies beneath your skin. It is sometimes loosely referred to as "belly fat" or "central obesity."

Today, white fat is recognized as an organ because it produces hormones and agents that cause inflammation.[161] There is emerging evidence that fat forms part of the body's innate immune system and it can become *activated* if the body senses it is under attack. When it is activated, it becomes inflamed and can release inflammatory agents that propagate the fire of inflammation.[162]

Activated and inflamed visceral fat may expand, and as it does so it releases even more inflammatory agents, which makes visceral fat a notable source of inflammation. Although visceral fat correlates with inflammation and activated visceral fat releases inflammatory agents, we are still not sure if visceral fat is the cause of inflammation or its consequence. It is likely both.[163]

There is also an association between intestinal permeability and visceral fat, and some scientists have proposed a theory that visceral fat becomes activated and may expand in response to "enemy" agents that leak into the body through a leaky gut.[164] The activated visceral fat releases inflammatory agents to help the body overpower the invaders. This theory places gut bacteria at the center of visceral fat dynamics.[165] [166] Inducing inflammation in the intestines of mice either with a high-fat diet or by causing colitis makes their gut leaky and activates visceral fat.[167] Giving the mice the probiotic *Lactobacillus gasseri* SBT2055 protects the intestinal wall from becoming permeable and prevents visceral fat from increasing.[168]

You are likely to have visceral fat if you are overweight or if you are thin but have a slight belly or a "spare tire."[169] Visceral fat can be measured in many gyms through bio-impedance measuring techniques. Although these may not be entirely accurate, a rough estimate can be helpful. Whether it plays the chicken or the egg in the saga of inflammation, it is good to carry as little visceral fat as possible.

If the theories regarding its relationship with a leaky gut are proven to be correct, then the best way to minimize visceral fat may be by looking after your gut bacteria and taking care that your intestinal walls are healthy and intact. Recovering faster from stressful experiences and being able to relax will also help.

There is some evidence that sugar in the diet may encourage the formation of visceral fat (perhaps by its effect on the microbiota) if cortisol levels are high, as might be the case in the setting of stress.[170] The study that demonstrated this relationship did not make a distinction between different sources of sugar, but this finding suggests it is wise to avoid all refined carbohydrates, particularly during times of stress.

Any approach that results in fat loss may help to reduce visceral fat.[171] [172] Regular aerobic exercise and restricting your calories can help.[173] If you lift weights in the gym, a study has shown that combining high-intensity training (two sessions a week) with regular gym training (two sessions a week) works better than regular training alone.[174] How often you eat might also make a difference. Eating six times a day may deposit more fat around the abdominal organs than eating three times a day.[175]

Fitness

As you improve your fitness, you are likely to improve overall levels of inflammation.

Exercise is good for inflammation, as long as you don't overdo it.[176] As you improve your fitness, you are likely to improve overall levels of inflammation. Physically active people have less inflammation than sedentary people.[177] [178]

Heat

This chapter has focused on reducing inflammation, which can, in a given context, raise the body's temperature. Intriguingly, raising the body's temperature *outside* the context of inflammation and illness can have a positive effect on mood. Regular sauna therapy improves symptoms of depression.[179] A randomized, controlled study has shown how raising the core body temperature to 100.4 °F (it is usually 99.3) on just *one* occasion rapidly improved the symptoms of fifteen patients suffering from major depression, compared to a control group. The effect lasted for at least six weeks afterward.[180] The thyroid hormone thyroxine can either swim freely in your blood or be transported around by a "carrier." A rise in body temperature makes the thyroxine detach from its carrier so that there is more thyroxine floating about. Artificially raising levels of this "free" thyroxine can improve symptoms of depression.[181] If the body temperature rises by 3.6 degrees from 98.6 °F to 102.2 °F, there is a 23 percent increase in the concentration of "free" thyroxine, which can induce euphoria.

Sauna therapy may benefit general brain health. Part of this effect may arise from the increase of what are known as heat-shock proteins (hsp), especially hsp70, through heat stress. Hsp70 proteins are thought to protect cells from oxidative stress.[182] A recent prospective population study conducted in Finland, in which more than two thousand healthy men were followed for two decades, revealed that sauna bathing can strikingly reduce the risk of dementia. The hazard ratio was reduced from 0.78 to 0.34 as the frequency of sauna baths increased from two to three sauna baths per week to four to seven sauna baths a week.[183]

- If you're feeling down, visit a sauna or take a hot bath.
- Try to regularly visit a sauna a few times a week (after checking with your physician).

Light

We associate sunlight with ultraviolet radiation, but more than half of all the sunlight we receive is in the form of *infrared* light.

When sunlight touches our skin, a third of it is in *near-infrared* form, which can travel past the skin's surface and penetrate into deeper tissue.[184] Near-infrared light may exert an anti-inflammatory effect.[185] Animal studies have shown it reduces levels of several inflammatory agents in inflamed sites of the body and it can "calm" microglia.[186] [187] [188] Applying near-infrared light to the head using a specially designed gadget may bring benefits to help reduce anxiety and improve memory retention.[189] In a double-blind, placebo-controlled study on forty healthy undergraduates recruited from the University of Texas, those who were given infrared light therapy to the head had faster reaction times and better recall performance after two weeks. Their mood also improved.[190] This is an emerging area of research and the findings are not yet definitive. Meanwhile, a relatively safe way of accessing near-infrared light is by tapping it from nature. Brief, safe, and regular episodes of sun exposure may benefit your vitamin D levels as well as harnessing the positive effects of near-infrared light.

Alcohol

Alcohol may increase intestinal permeability, promoting inflammation. One study has elucidated how this might happen. Alcohol appears to target the mucus layer that lines the intestinal walls; it extracts lipids from within the layer and reduces its hydrophobicity. Its effect on gut permeability appears to be dose-dependent.[191] You might recall that soluble fiber has a protective effect on the mucus layer, and I described a mouse experiment earlier in which soluble fiber prevented mice from developing alcohol-induced gut leakiness. This experiment has not been replicated in humans, and given its potent inflammatory effect, I would *drastically reduce* alcohol if you are aiming to repair your brain and gut from the ravages of stress. Alcohol consumption has recently been declared a cause of at least seven cancers, with no amount deemed entirely safe, so you may be doing your entire body a great favor by avoiding it altogether.[192]

CHAPTER 7

Modulating
Insulin Resistance

IF YOU FORCE an unsuspecting little mouse to share its home
with a mouse whom it doesn't quite warm to, it will become stressed. If the other mouse makes it feel less important,
the poor mouse develops "subordination stress." If you were to
now peer into its liver and muscles, you would discover that as
the mouse becomes more and more stressed, it also becomes *insulin resistant.*

Insulin resistance is one of our agents of stress. If you are
about to be mauled by a lion, you need your brain to coordinate
an escape strategy. Its needs outweigh those of the rest of your
body, so your body presses a switch that stops the glucose in your
blood from moving into your body's cells. As glucose piles up in
your blood, more is able to reach your brain.

The clever switch that your body presses in order to make this
happen is insulin resistance. A set of doors separates your body's
cells from your blood. These doors need to be unlocked in order
for glucose to gain entry. Insulin carries the keys. It unlocks the

doors all day long, so that every time the level of glucose in your blood rises, some can be pushed out of your blood and into your body's cells, bringing the level in the blood back down. This keeps glucose levels stable. If insulin ever loses its keys, it can't unlock the doors and glucose can't move out of your blood and into your body's cells. It gathers in the blood and its level rises. When insulin loses its keys, we call it insulin resistance. In the context of a short and sweet stress response, it can save your life. If insulin resistance overstays its welcome and becomes your perpetual state, then instead of saving your life, it may shorten it.

Insulin resistance often occurs hand in hand with central obesity (fat accumulation around the middle) and high blood pressure. When they happen together, they are referred to collectively as *metabolic syndrome*, partly because they are a consequence of metabolism going awry.

You might know of people who used to be slim and fit at the start of adulthood, but have since led stressful lives and lost their vigor at a rate that is disproportionate to normal aging. They may now be overweight, they may be carrying excess fat around their middle—what is sometimes referred to as middle-aged spread—and they may be under surveillance from their physician because their blood sugar levels and blood pressure are rising and their cholesterol levels are not looking good. In parallel with these changes, they may be experiencing subtle changes in mental clarity, mood, and even mental performance. These may be early signs of metabolic syndrome.

There is an emerging link between chronic stress and insulin resistance. Chronic stress has been shown to bring on a state of insulin resistance in rats with as little as a few hours of stress exposure every day for two weeks.[1] Due to ethical constraints, it is difficult to conduct a rigorous experiment in the laboratory testing chronic stress and insulin resistance in humans. Instead, we can look at groups of people in real life, map their stress levels, and see if there is a correlation with insulin resistance.

When 234 police officers were followed over a five-year period, those with higher stress levels were at an increased risk of developing metabolic syndrome.[2] A study on 1,815 male workers in Japan has shown how those with diminishing supervisor support had a significantly higher risk of having insulin resistance.[3] One of the contributors to workplace stress and burnout is the feeling that your hard work is not being rewarded. This feeling can be quantified as the **effort reward imbalance ratio (ERI)**. The larger the ERI, the less perceived reward there is for effort. A study on 1,441 German workers found a positive association between the ERI score and suffering from metabolic syndrome. The association was stronger for younger employees and for male workers.[4] A study on 204 Jordanian male workers revealed that if workers were under stress while having a high ERI score, they were far more likely to develop metabolic syndrome. If they had a high ERI score and were also inflamed, their risk increased even more.[5]

Hostility and anger not only hurt the mind— they also hurt the body.

Hostility and anger not only hurt the mind—they also hurt the body. The tendency toward anger and hostility is associated with insulin resistance, independent of other factors such as obesity, and the association may be stronger in some more than in others.[6] One study looking at healthy middle-aged men has shown how those who experienced "hostile paranoia" and "vital exhaustion" also tended to have insulin resistance and higher levels of inflammation.[7] How anger or hostility in your mind can directly influence the ability of insulin to open the locked gates in your muscles is not yet fully understood; some studies suggest the two separate phenomena are connected by serotonin signaling within the brain, which plays a role in the release of the stress hormone cortisol, which in turn influences blood sugar control.[8]

ⓔ WHY IS INSULIN RESISTANCE SO DAMAGING? ›

There are two factors at play in the setting of insulin resistance, glucose and insulin. Both are vitally important for survival, but they work their magic within a narrow bandwidth. If their levels are too high or too low, the brain and body suffer.

Glucose and the Brain

Insulin resistance can raise blood glucose levels. Disturbances in glucose levels, in either direction, can harm the brain through many avenues.[9]

Raised glucose levels can increase the risk of glutamate toxicity.[10] [11] You have come across AGEs in the context of inflammation. AGEs form when glucose reacts with fat or protein, and may be present in certain types of food. AGEs can also form in the setting of raised blood glucose and may cause harm to the brain.[12] [13]

If enough glucose is not available to the brain, it runs out of energy and can't function optimally. This is why, if your brain detects falling levels of sugar in your blood, it triggers a stress response. Stress transiently increases glucose levels in your blood in order to provide your brain with an oversupply of energy while you are under attack.

Synaptic plasticity is profoundly expensive. It is heavily dependent on energy supply. If you deprive brain cells of energy, you shackle their freedom to form synapses in any way they choose.[14] In Alzheimer's disease, the brain resorts to a "cost-effective" pattern of connectivity that compromises function in return for lower energy usage, just like a company undergoes restructuring to make savings. One theory is that this happens because the brain struggles with getting enough glucose.[15]

You might think that eating a diet that is high in refined sugar is good for providing the brain with energy, but this may have the *opposite* effect. In a Japanese study, twelve perfectly healthy young men and women were given a sugary drink (containing 75mg of glucose) after not eating for eight hours.[16] Their brains were then examined under a scanner. Rather alarmingly, as the level of glucose in their blood rose, parts of their brains seemed to *raise the drawbridge* to glucose. Glucose could no longer enter their cells to provide energy. The affected areas included parts of the prefrontal cortex. This phenomenon is known as glucose hypometabolism, and it has been linked to depression.[17] In a study on depressed older patients, the more sugar there was in the blood, the greater was the reduction in sugar uptake by the brain.[18] Insulin resistance may induce brain glucose hypometabolism.[19]

Insulin and the Brain

Insulin's actions and dynamics become compromised in the setting of insulin resistance. This is terrible news for a stressed brain because insulin *encourages synaptic plasticity*. Injecting the hippocampus of healthy rats with insulin *improves* learning and memory.[20] Giving insulin through the nose to mice sharpens both short- and long-term memory.[21] Giving insulin in this way to humans *also* improves memory.[22]

So if chronic stress can lead to insulin resistance, is there anything we can do? Although chronic stress increases the risk of insulin resistance, that risk is further increased in the setting of several other factors that can independently raise the risk of insulin resistance. If you can eliminate these other factors, you might be able to protect yourself from insulin resistance even if you are suffering from chronic stress.

☺ RISK FACTORS FOR INSULIN RESISTANCE

Inflammation is associated with insulin resistance, so please refer to chapter 6 for more information on inflammation. Aside from inflammation, dietary patterns, sedentary behavior, visceral fat, and irregular sleeping patterns are all associated with insulin resistance.

Dietary Patterns

Front-loading your meals may reduce your risk of insulin resistance. In one study, people with insulin resistance improved glucose and insulin control significantly by eating 700 calories in the morning and 200 calories in the evening instead of 200 calories in the morning and 700 calories in the evening (and a constant number of calories at lunchtime).[23] In another study, skipping breakfast and fasting until noon impaired insulin dynamics.[24]

There is something known as dietary acid load, which refers to the portion of your diet that shifts the pH of your blood toward acidity and hence requires more buffering by your kidneys. Common foods that may have a higher acid load include meat, cheese, eggs, and cereal grains, whereas fruits and vegetables have an alkalizing effect.[25] There is some suggestion that having a high dietary acid load is associated with insulin resistance.[26]

FAT

- Avoid excessive palmitic acid

    A high-saturated-fat diet (65 percent) composed of 40 percent palmitic acid rapidly gives rise to insulin resistance, when compared with a low-fat diet in which palmitic acid contributes only 8 percent of the total calories, even if the low-fat diet is based on muffins, pie filling, and ice cream.[27] Sources of the long-chain saturated fat palmitate include animal and dairy fats, such as butter, lard, tallow, and cheese.

- Don't saturate your meals with fat.

    There is some emerging evidence that regularly eating a large portion of fat may affect the production of bile acid, which can affect the intestinal environment and gut bacteria, which may then raise the risk of insulin resistance.[28] [29] Different fats seem to have different effects. One small study, for instance, has demonstrated that medium-chain saturated fats have a smaller effect on raising levels of bile acids in the blood than other fats.[30] This is, however, a developing story, and at this stage we can only watch and wait. More on dietary fat later.

SUGAR

- Avoid *any* form of processed sugar.

    Consuming sources of refined carbohydrates, including bottled juices, soft drinks, and added sugar of any kind is associated with insulin resistance.[31] Fructose forms half of a sucrose molecule, and fructose is often added to various types of processed food and drinks. It has the ability to induce insulin resistance in the liver independently of insulin and glucose, and may be contributing to the link between processed food consumption and insulin resistance.[32] The combination of simple sugar and the saturated fat palmitate may be particularly bad for insulin resistance.[33] [34] [35]

■ Choose your carbohydrates carefully.

A Japanese cohort study has shown that if your BMI is 25kg/m² or more, carbohydrates should not contribute more than 50 percent of your total calories.[36] If you have some degree of insulin resistance, then reducing your carbohydrate intake at every meal to no more than 30 percent of the total calories (while eating no more total calories in a day than your body needs) may help your insulin dynamics.[37] In general, it's best to avoid refined, processed carbohydrates. At the very least, avoiding refined carbohydrates in the evening may protect you from the risk of insulin resistance as long as your overall diet pattern resembles the Mediterranean diet.[38] If you choose to eat wheat, then dense, non-leavened breads such as flatbreads are better for your insulin dynamics than fluffy risen bread, and rye bread may be better than wheat.[39] [40]

If you are both stressed *and* have any semblance of gastrointestinal symptoms, please refer to chapter 6 for information relating to IBS.

■ Choose Low-GI Starches

With regard to insulin dynamics, starchy food is better than refined sugar, but slow-digested starch is better than fast-digested starch. One example of slow-digested starch is green bananas. Cooling certain items after cooking (boiled potatoes or white rice), by storing at 39 °F for twenty-four hours, can also increase their "resistant-starch" content.[41]

One study has shown how eating white rice with soluble fiber added to it can *improve* insulin resistance in as little as two weeks, perhaps because the soluble fiber reduces its rate of absorption, turning it into a "slow-digested" starch (more on this below).

PROTEIN

The World Health Organization and the Food and Agriculture Organization recommend that a nonpregnant, nonathlete adult consume 0.8g of protein per kilogram of body weight per day.[42] Some studies on protein suggest there may be a link between excessive protein consumption and insulin resistance, although the context and source of protein may be important. One study has found a direct association between total and animal protein intake, but not plant protein intake, and insulin resistance, while another, on older men and women, found an indirect association between animal protein and insulin resistance.[43] [44] When obese postmenopausal women were given two whey protein shakes per day as part of a calorie-controlled weight loss diet, they did not benefit from the improvements in insulin dynamics that one would expect with weight loss.[45] [46] Meat intake, especially processed meat intake, may be linked to insulin resistance.[47] Interestingly, fish protein (from sardines) has been shown to be beneficial for insulin resistance and glucose control in diabetic rats.[48]

- Avoid *all* refined sugar.
- Choose whole forms of grain over flour.
- If you don't have IBS, rye bread may be better for insulin dynamics than wheat bread.
- Dense bread may be better for insulin dynamics than light, fluffy bread.

- Eat white rice together with soluble fiber or after refrigerating for twenty-four hours after cooking.
- Unless you are an athlete, avoid processed protein powders.
- If your meal is heavy in protein, replace as much animal protein as possible with plant protein sources such as legumes.
- Balance a protein-heavy plate with fresh vegetables.

AGEs

As we have seen in chapter 6, AGEs can increase inflammation. AGEs in the diet may also raise the risk of insulin resistance.[49] Animal products that are high in fat and protein may contain significant amounts of AGEs, especially after they are cooked. Roasting nuts, grilling steak, sautéeing tofu, and frying meat (including bacon) may increase AGE content. Breaded chicken or fish has high levels of AGEs. Roasting, then barbecuing meats markedly increases AGEs. Scrambled eggs may have a lighter AGE load than other animal proteins. Boiled white rice is particularly low in AGEs. Vegetables, fruits, milk, and whole grains have lower levels of AGEs than animal proteins even after they have been cooked, if they are not cooked with added fat. Cooking with moist heat may give rise to fewer AGEs than using dry heat.

FIBER

One Japanese study has demonstrated that adding soluble fiber to white rice changes its effect on blood sugar and insulin. Eating a breakfast containing white rice and soluble fiber for just two weeks actively *improved* insulin resistance in a group of overweight men and women when they switched from eating an identical breakfast that did not contain the soluble fiber.[50] Another study found a similar result.[51] Adding 6g of psyllium husks to a carbohydrate-based breakfast improves insulin dynamics following the meal.[52]

FERMENTED MILK

Several studies from around the world have shown an inverse relation between regularly eating a fermented milk product such as yogurt or kefir and insulin resistance.[53] [54] Make sure the yogurt is unprocessed and the label lists the bacteria it contains.[55] Aim for at least 300g a day.[56] If you take a course of antibiotics, use a short course of a high-quality probiotic supplement afterward to replenish your gut's ecosystem, then maintain it daily with yogurt. The probiotic VSL#3 has been shown to reduce the risk of autoimmune diabetes.[57]

- Eat 300g of natural, plain, probiotic yogurt every day.
- If you need to take a course of antibiotics, consider replenishing your gut microbiota with a short course of a probiotic formulation such as VSL#3 (after checking with your physician).

TURMERIC

In addition to its anti-inflammatory benefits, turmeric may also help protect against insulin resistance.[58] There is conflicting information on exactly how much should be taken to reap its benefits, and how much of its active ingredient, curcumin, actually gets absorbed when taken as a pill is a matter of debate. For this reason, I prefer to take the whole spice and try to emulate what is done in India, where it is added to almost every meal as part of the cooking process, including breakfast.

Turmeric is "broken down" by the liver, and this process may be decelerated by the compound piperine, found in black pepper.[59] Eating curcumin with a pinch of black pepper drastically raises its level in the blood an hour or so after ingestion.[60] Mixing turmeric into fat (in effect, eating turmeric as part of a meal that contains fat) may also increase its bioavailability. Finally, it is better to take the whole spice rather than a pill containing extracted curcumin because other compounds present within whole turmeric seem to improve its absorption and retention, and also because taking the whole spice maintains some degree of dilution that might protect from toxicity.[61][62] If you have biliary tract obstruction, are pregnant, or have been told to limit your oxalate intake, please check with your physician before adding turmeric to your current diet.

- If you take turmeric, choose the whole spice if possible, rather than a pill containing an "extract." Consider adding it to your cooking. If you take a pill, aim for one that contains whole turmeric.
- Take turmeric with a quarter teaspoon of black pepper and with a little fat.

CINNAMON

Although the common kitchen spice cinnamon improves insulin resistance in rats, this effect has not consistently been found in humans. Where studies have shown some degree of benefit, the dose of cinnamon used was between 1g and 3g daily.[63][64] Please check with your physician before adding large amounts of cinnamon to your diet.

MAGNESIUM

A meta-analysis of eighteen randomized trials has shown that magnesium supplementation can improve insulin dynamics in people who are at risk of developing diabetes, such as those who are overweight or obese.[65] A broad spectrum of doses and magnesium compounds were included in the analysis. You can improve your magnesium intake naturally through your diet by consuming more dark green vegetables, seeds, and nuts. Pumpkin seeds, sunflower seeds, almonds, and cashews all contain good amounts of magnesium.[66]

◎ NON-DIET RISK FACTORS FOR INSULIN RESISTANCE

Other aspects of lifestyle that affect your risk of insulin resistance include:

- Sedentary behavior
- Having visceral fat
- Irregular sleeping patterns

Sedentary Behavior

A study on young and healthy men and women has shown that if you spend too long sitting down *and* you are eating more calories than you need, insulin is 39 percent less able to do its job, whereas if you are eating no more calories than you need, it is only 19 percent less able. Your body's resistance to insulin rises by 19 percent simply on account of not moving, even if you are burning all the energy that you are taking in. Physical activity *on its own* reduces the risk of insulin resistance regardless of diet. If your diet is also poor, then sedentary behavior acts *in synergy* with it to cause insulin resistance. The less active and fit you are, the more benefits you will reap from moving around.

> *The less active and fit you are, the more benefits you will reap from moving around.*

Just one day spent sitting for over sixteen hours can make the entire body insulin resistant, when compared to a day spent sitting for only six hours, even in people who are healthy and young.[67]

Going for a walk for fifteen to forty minutes immediately after a meal, instead of sitting down, significantly improves insulin dynamics.[68] [69] The longer you can walk, the better. Walking for five minutes or doing simple resistance exercises such as squats and calf raises for three minutes every half hour also improves glucose and insulin levels in the setting of insulin resistance.[70] [71]

One study has shown it may be better for your insulin dynamics to slot in several short bouts of walking at a moderate speed rather than to sit all day and then spend thirty minutes at the gym.[72] Ideally, you want to *both* incorporate short bouts of movement during your day *and* fit in some dedicated exercise time.

If you have insulin resistance, it may be better to exercise a short while *after* eating something rather than *before* you eat. Exercising while you are in a fasted state, *before* a meal, may impair glucose control after the meal, especially if the exercise is of a high intensity.[73] If you are sedentary and have insulin resistance, *low-intensity* exercise (such as cycling for sixty minutes at 35 percent of maximal effort) may improve insulin control for the remainder of the day, to a greater extent than higher-intensity exercise (such as cycling for thirty minutes at 70 percent of maximal effort).[74]

A study on sedentary men who had not been diagnosed with insulin resistance found that exercising intensely at all-out effort for ten minutes, three times a week for three months resulted in similar improvements to insulin resistance as exercising at 70 percent of maximal effort for fifty minutes, three times a week for three months. The ten-minute protocol consisted of three twenty-second intense cycle sprints interspersed with two minutes of very light cycling, with two minutes spent warming up and three minutes spent cooling down.[75]

- Take a walk after *every* meal, for at least fifteen minutes.
- If you have a sedentary job, stand up and walk around your office every thirty minutes.
- Alternate walking around with doing a few squats, push-ups, jumping jacks, squat thrusts, or jogging in place for three minutes.
- Minimize your seated time. Schedule walking meetings and phone calls. Walk instead of using public transportation.
- If you are sedentary for a defined period of time, *never* eat more calories than your body needs around that time period.
- Even if you manage to incorporate some walking into your day, squeeze in some dedicated exercise time.

Visceral Fat

Visceral fat and insulin resistance are inextricably linked. If you have one, you are at risk of having the other. For more information on reducing visceral fat, please refer to chapter 6.

Irregular Sleeping Patterns

Just two days of sleeping from 2:45 A.M. until 7 A.M. disrupts insulin dynamics, compared with sleeping from 10:30 P.M. until 7 A.M.[76]

Insulin sensitivity is best during the "active" hours in mammals, the daytime in humans, and it is best to restrict eating to the daytime for this reason.

■ Try to sleep at least seven hours every night.
■ Try to limit eating from evening onward.

◎ NUTRITION FOR THE BRAIN

After you have adjusted your diet and lifestyle to minimize the risk of insulin resistance, you can reinforce your brain's resilience further with some nutritional interventions.

Ketones

If you are working away intensely and start feeling hungry and mentally exhausted because your blood sugar is falling, your cognitive performance may start to slide.[77] In one experiment, some diabetic volunteers were given insulin, which made their blood sugar sink very low. As expected, their cognitive performance suffered. Intriguingly, if they were also given some medium-length saturated fats, known as MCTs, their cognitive performance did not plummet.[78]

Your brain usually uses only glucose as fuel but it is also able to use something known as a ketone. The body turns fat into ketones to provide the brain with fuel when its glucose supply runs thin. Early in vitro findings suggest that while other MCTs get turned into ketones by the liver, the fat lauric acid (which is the predominant fat in coconut oil) may be converted into ketones by astrocytes in the brain. Although coconut oil may not raise the level of ketones in your blood by much, it may substantially raise their level within the brain in this way.[79] When people suffer from glucose hypometabolism and glucose cannot enter into certain parts of the brain, the entry of ketones seems to be unaffected. One study on patients with Alzheimer's dementia found that 1.35 ounces per day of virgin coconut oil resulted in a statistically significant increase in cognitive test scores after twenty-one days.[80] A study done on mice has shown that a diet in which 30 percent of the total calories consumed came from ketones increased the speed of mental processing by 38 percent and improved mental performance.[81]

You can provide your brain with ketones intermittently by exercising, including a long gap between your evening meal and breakfast the following morning, and never eating more than your caloric needs. Including some coconut oil in your diet may, in theory, also supply your brain with ketones. Although coconut oil can affect cholesterol levels, some studies suggest virgin coconut oil may be less harmful to cardiovascular health than previously assumed.[82] [83]

- If you are considering adding coconut oil to your diet, make sure you use virgin, unprocessed coconut oil. Start with a small amount every day, such as a tablespoon (after checking with your physician). Cooking with virgin coconut oil in place of your usual cooking oil is one way to incorporate it into your diet.

Fish Oil

Docosahexaenoic acid, or DHA, and eicosapentaenoic acid, or EPA, are derived from fatty fish and are collectively referred to as omega-3 fish oil.

A placebo-controlled, double-blind, twelve-week, random-ized, controlled trial on sixty-eight healthy young medical stu-dents found that taking 2.5g or 0.088 ounces of fish oil per day (amounting to 2085mg of EPA and 348mg of DHA) reduced anxiety symptoms by 20 percent.[84] Giving twelve healthy young adults 750mg of DHA and 930mg of EPA per day for six months improved their working memory.[85] Finally, a daily dose of 1.5g DHA and 360g EPA for six weeks reduced perceived stress in staff at an Australian university.[86]

It may be best to consume fish oil from eating whole fish rather than fish oil supplements. When I use a supplement, I fa-vor a cold-extracted whole-source fish oil that contains omega-3 as well as the other ingredients that normally "package" the omega-3 fat when it is inside the fish, in the hope that this minimizes the risk of oxidation. Fish oil supplements are absorbed best when taken with food that contains fat.[87] [88] [89] [90]

The Question of HDL Cholesterol

The cholesterol in your blood is packaged in many different ways, and a glimpse into the "pattern" of packaging can give clues about your overall health and risk of disease. The packages include acronyms such as LDL-c, LDL-P, CLDL-c, non-LDL-c, HDL-c, and HDL-p. We tend to associate our blood cholesterol levels with the risk of heart disease, but one of these markers, HDL-cholesterol, or HDL-c, may also relate to stress resilience.

There is some evidence that depression may be associated with lower levels of HDL-c.[91] [92] When people are successfully treated for their depression, their blood cholesterol levels can rise.[93] In one study, twenty healthy young volunteers were made to follow a high-fat diet for one month, then half switched to a low-fat diet. The low-fat eaters reported feeling more tense and anxious than before. Blood tests revealed their HDL levels had fallen. The high-fat eaters felt *less* tense and anxious and their HDL levels rose.[94]

Dietary fiber is associated with higher HDL-c levels, particularly in the setting of insulin resistance.[95] Removing all refined carbohydrates from your diet and replacing a "typical American diet" with a "Mediterranean-style" diet, in which butter and animal fat are replaced with fat from olive oil and avocados, can raise HDL-c levels.[96][97] A study from Spain showed that extra virgin coconut oil raises HDL-c levels without purported negative effects if the diet is otherwise healthy.[98] The main fat within coconut oil, lauric acid, can raise levels of both HDL and LDL cholesterol, but its effect on HDL-c may be greater.[99][100][101]

- Eat more dietary fiber.
- Replace butter with extra virgin olive oil.

You should not actively aim to raise your HDL-c if this is within a healthy, normal range, because an excessively high level of HDL-c may signal harm. Please check with your physician before making any changes to your diet.

Vitamins

B VITAMINS

Only three months of taking B vitamins significantly reduces the stress felt at work.[102] They work against depression and mental fatigue, and they improve mood and cognitive performance.[103][104][105][106] If you buy them as a vitamin supplement, they are often sold as a set, called B vitamin complex. Make sure the complex includes B12.[107][108] *B vitamins may work well if you are also getting a good supply of omega-3 fish oil.*[109] One seems to need the other. They work as a team.

VITAMIN D

Vitamin D has many roles in brain health and may improve memory.[110 111 112 113 114 115 116 117] In a small trial that used a lifestyle-based approach to reverse memory loss in a small group of people with mild cognitive impairment, taking 2,000 IU (international units) of vitamin D per day was recommended to several participants as part of the protocol.[118]

Increased vitamin D levels in the blood correlate with:

* reduced depression
* the ability to tolerate mental stress
* improved cognition

Vitamin D is acquired through sun exposure. Using a sunscreen with an SPF of 15 can reduce your skin's ability to produce vitamin D by approximately 95 percent, so if you are sunbathing for vitamin D it may be best to expose clean skin to the sun for a short period of time, either early in the day or late in the day, depending on where you live, so there is no risk of sunburn.[119] It is wise to get your vitamin D formally tested (as serum 25-hydroxyvitamin D) and to aim for the mid-range to upper-mid-range level of normal.

If you can't replenish your vitamin D levels with sunshine, you may be able to acquire vitamin D naturally from your food. Dairy products and fish contain vitamin D, but the amounts can be variable. One of the densest dietary sources of vitamin D is the dish known as *mølje*, consumed in northern Norway during the winter months. Mølje is made from cod liver and cod roe, and northern Norwegians are thought to be protected from the higher incidence of multiple sclerosis found in the south of Norway, by eating mølje.

Vitamin D is a fat-soluble vitamin and is best taken with a meal that contains fat.

- Check that your vitamin D status is at the mid-range to upper-mid-range of normal.
- If it isn't, try to increase sun exposure, safely.
- If you don't have much access to sunshine, and a blood test suggests your vitamin D levels may be low, consider a supplement of 1,000 to 2,000 IU of vitamin D (after checking with your physician).

CHAPTER 8

Mastering Motivation

IF YOU ARE being chased by a lion, you feel motivated to run for your life. If, instead, you were to shrug and decide you couldn't be bothered to run, you would likely meet an unfortunate end. An acute stress response momentarily increases your motivation so there is no risk of your surrendering to the situation without a fight. Motivation operates by tempting you with a reward that you desire enough to act on. In the case of the lion, your reward is to escape from the jaws of death. You act on it by running away. If you manage to escape, you learn a lesson for next time— that it really is possible to escape being eaten by the lion (the reward) by running at top speed (the action), and this removes your doubts and increases your motivation to run the next time you see a lion. Although acute stress increases motivation, chronic stress can make motivation go awry.

A system in your brain known as the reward circuit acts to give you an incentive for doing things. There are many different ideas regarding the precise mechanisms behind reward and motivation, but one school of thought is that motivation for a reward has three components: *wanting, liking,* and *learning.*

- Wanting: When you *anticipate* pleasure, your reward circuit fires and motivates you to reach for it.
- Liking: When a thing or a person or a situation makes you *feel* pleasure, your reward circuit fires.
- Learning: You learn that performing an action leads to feeling pleasure and you know that to perform this action again will give pleasure.

When scientists want to induce depression in mice they subject them to two forms of chronic stress, which resemble the kinds of psychosocial stress that we humans suffer from in our day-to-day lives and can also bring on depression in us.[1] Both of these models of stress bring on *anhedonia.* Anhedonia is what happens when you no longer feel pleasure from doing things you would usually enjoy and are not drawn toward activities that will bring you pleasure. *Acute* stress increases your motivation. *Chronic* stress can bring on *anhedonia.*

"LEARNED HELPLESSNESS"

Learned helplessness is the process of learning that you have no control over your situation. No matter how unpleasant it is, there is nothing you can do to make things better or to escape from it. The mice are subjected to an incessant series of small electric shocks that they cannot escape from. At first, they actively try to find a way to escape. Eventually they become depressed and give up. If you then offer those mice an escape route to run away from the foot shocks, they won't take it. They have almost become numb to pain and are no longer motivated to seek a respite from pain. They develop anhedonia.

☺ CHRONIC SOCIAL DEFEAT

Social defeat is the experience of being subjected to bullying, aggression, and social conflict. When a vulnerable mouse is locked in a cage with an aggressive mouse for about ten minutes every day, it develops depression after only two weeks. Ten minutes of mouse time is at least as long as a workday in human time. The mouse becomes socially withdrawn and develops anhedonia.

Anhedonia results from a defect in the reward circuit. The reward circuit is likely affected through several different channels in chronic stress, including the channels of inflammation, a dysregulated body clock, and an excess of stress hormones. Networks within your prefrontal cortex engage in dialogue with networks in your reward circuit, supported by networks in various other parts of your brain, including those related to emotion processing, as they help to plan your behavior with a goal in mind. As chronic stress changes the brain, this complex conversation may suffer.[2]

A healthy, normal state of mind usually functions in three modes: a baseline, a positive mode (such as joy), and a negative mode (such as sadness). The baseline state is neutral and is interrupted with segments of negatives and positives. We cannot experience profound joy *all* of the time but our existence is peppered with it, as it is peppered with deep sadness. The positive modes involving pleasure and joy are brought about by the reward circuit. It tempts us toward pleasure and enables us to enjoy it.

When people suffer from depression with anhedonia (like the

mice described above) the mind operates in a negative mode *more* often, and it operates in a positive mode *less* often. A dysfunction in the reward system is thought to be part of the reason for operating in the positive mode less often. The negative mode can be controlled with techniques like emotional regulation; however, rescuing the mind from a negative mode does not necessarily place it into a positive one. A separate and dedicated approach may be needed to rekindle positive emotions in a chronically stressed mind.

Anhedonia is very challenging to treat. Since it may arise from a malfunctioning reward circuit, many current approaches to anhedonia focus on stimulating the reward circuit. One such approach, "Positive Affect Treatment," proposed by two teams at University of California, Los Angeles, and Southern Methodist University, Dallas, involves a three-step method that focuses on training *wanting, liking,* and *learning.*[3] Briefly, the first module, "pleasant events scheduling," encourages clients to plan events that bring pleasure. This planning process makes the clients *want* or *anticipate* the event and this anticipation activates the reward circuit. There is some evidence that positive emotions can be intensified with reinforcement and the clients are trained to savor and dwell on the pleasure that the planned events will bring. The second module involves cognitive training and includes techniques for training clients to find and appreciate patches of pleasure in everyday situations and to identify the behaviors that lead to pleasure (*learning*). The third module focuses on *liking* through giving and gratitude.

This technique has not been tested on chronic stress, but there is evidence that actively cultivating reward and pleasure in your daily life can protect you from chronic stress and its damaging effects on your reward circuit. As you become increasingly exhausted and time-poor, you may start shaving off commitments that you deem dispensable. Often, "non-essential" things, such as the things that you do purely for pleasure, are the first to go. If, when you fall under the daily grind of a stressful life, your *want* wanes, your obstacles will block your motivation. If you stop doing things that bring you pleasure, while living under chronic stress with a hyperactive emotional brain that keeps you immersed in negativity, your life will descend into a spiral of gloom.

You must treat pleasure with the same importance with which you treat going to work or taking a shower, by allotting time for it every day. Never leave pleasure by the wayside or sacrifice it for "more important" things.

Never leave pleasure by the wayside or sacrifice it for "more important" things.

THE POWER OF REWARD

Musical-theater songwriter Robert Sherman came home one day to learn that his son had just been given his polio vaccine. When he asked if it had hurt, his son replied that no, the doctor had placed the "medicine" on a spoonful of sugar. That is how the famous Mary Poppins song "A Spoonful of Sugar" was born. Robert Sherman wrote the lyric, his brother the tune. In mice, this "spoonful of sugar" may reverse the negative effects of a stressful ordeal, if given soon after the ordeal, as a reward for effort. In one study, one half of a group of happy mice was subjected to social defeat stress for four weeks. After the four-week period, all of the mice were taught new spatial tasks and their ability to navigate their way around a new challenge was tested.

The stressed mice appeared unmotivated and fared much worse on the tasks than the unstressed mice, with an interesting exception. Some of the stressed mice were given a sugar reward as they were taught the spatial tasks. This reward appeared to "normalize" their behavior so it was comparable to that of the mice who had not been stressed.[4]

This experiment showcases the profound importance of feeling pleasure and its power to grant you stress resilience.

◎ REWARD EXPERIENCE

A study carried out on almost five hundred female twins (to rule out any genetic effects) has shown that actively seeking and feeling pleasure at every opportunity in your daily routine can make you resilient to stress.[5] You have a "pleasure piggy bank." You need to fill it with as many pleasure coins as you can, every day. The total you collect is known as your **reward experience**. Your reward experience confers stress resilience, no matter how stressful your day has been.

In light of these observations, you must actively try to incorporate moments of pleasure within your day and find patches of pleasure in your field of life. Some strategies to increase or incorporate pleasure are:

- Forming new "pleasure habits"
- Using the "closure principle"
- Being mindful
- Laughter
- Using music
- Using gratitude

Forming New "Pleasure Habits"

If you have shaved so much off your pleasurable time that you no longer do anything just for the sake of pleasure, your "reward experience" will suffer. If you want to protect your brain, you need to start reincorporating habits that lead to pleasure. You need to create *new habits* that do not relent to apathy or laziness.

There is a form of therapy used in depression known as "Engage," in which the therapist helps the client to choose rewarding activities, identify the steps needed to experience each one, and create action plans with clear steps as well as a backup contingency plan. The therapist works through the client's reluctance to engage in pleasure. Placing new cues to form new habits and removing old cues to forget old ones, as well as setting reminders, all help to resist apathy.

You can use the principle of Engage in the following way. Pick three things that bring you pleasure. Look at your schedule and slot these three things into your week. Let's say one of your activities is a boxing lesson you want to attend tomorrow.

- **Make an action plan checklist.** No matter how unnecessary this seems, make a checklist that requires you to check boxes.
  1. Book the reservation.
  2. Take out boxing gloves.
  3. Pack boxing clothes.

- **Create new cues.** A *new* habit needs a *new* cue.
  1. Hang your boxing gloves next to your television so they cheekily continue to remind you of your session.
  2. Enter reminders into your smartphone so it beeps reminders.

3. Place your gym clothes on the sofa so you are forced to pick them up whenever you are tempted to sit on it.

- **Remove old cues.** Old cues will push you into old habits.

  If you are used to coming home and slouching in front of the television all evening, fill your sofa with obstructions so you can't sit on it. By *removing* the cue of sitting on the sofa, you are remedying your old habit.

If during the next day you think of reasons why you don't need to attend your boxing session, write these down in a "negative thought logbook." After each reason, write two counterarguments. Once you have spent a couple of weeks doing this, you will form a habit.[6]

- Decide on three things you will do purely for the sake of pleasure.
- Pencil them into your schedule—with priority.
- Write an action checklist.
- Set new cues.
- Remove old cues.

You could also incorporate the wisdom from the first module of the "Positive Affect Treatment" approach I described earlier by dwelling on the pleasure that the events are likely to bring you and reinforcing the positive emotions you feel, as much as you can.

The "Closure Principle"

Completion triggers our reward circuit. We are programmed to want to resolve an unstable or an unfinished situation as soon as possible. When we do, we feel pleasure.

KEEPING YOUR REWARD CIRCUIT ACTIVE AS YOU WORK

You can use the closure principle to both stimulate your reward circuits *and* increase your motivation while you work. If a task is enormous and feels impossible to complete, the potential for pleasure from completing it will not motivate you. The prospect of feeling rewarded along the way will seem bleak. What you can do in this situation is bring the pleasure of completion closer, so it feels attainable enough to tempt you. You can do this by breaking it into smaller parts. As you complete each small task, your reward circuits fire and your motivation climbs. In his book *Getting Things Done*, the productivity expert David Allen recommends making the first small task last only two minutes because you won't want to procrastinate on something that can be done in two minutes. It makes getting your first pleasure kick easy. Motivation is like momentum. As it adds up, so will your drive to continue with the project. Your motivation to get the job done increases the nearer you get to completing it.[7] Another way to use the completion principle to keep yourself motivated is to *never stop at the end of a task*. Stop right after you have just begun a new task so that it remains annoyingly incomplete and you feel an urgent need to complete it when you return to your computer.

■ *Break up every task or activity into small components, and treat yourself to a small reward as you complete each one.*

CHOOSING ENTERTAINMENT THAT EXCITES
THE REWARD CIRCUIT

Thrillers, soccer matches, detective novels, and soap operas take us through a slalom of plot twists, surprises, and unpredictability as we long for completion. A theory known as the

excitation-transfer paradigm posits that "the intensity of the pleasure experienced during the resolution depends on the intensity of the (negative) tension experienced prior to the resolution."[8] Complex classical music can trigger reward circuits for a similar reason. The repertoire takes you on a journey with unpredictable syncopations and tortuous undulations, and when the piece finally ends you feel satisfied.

- *Immerse yourself in a thriller or a gripping detective novel on a regular basis.*
- *Watching a ball game is another good option!*

Being Mindful

Being mindful means paying attention to things around you rather than carrying on in autopilot mode. If you are attentive to your surroundings, you are more likely to spot things that bring you a moment of enjoyment. If you pay attention to your internal state, you may find you sometimes experience pleasure without noticing it. For instance, you might drink a cup of coffee while studying your computer screen at work and its pleasant taste evades your attention. If instead you take five minutes to look at your coffee and savor its taste and smell, that will bring you five minutes of pleasure.

- *Disengage your autopilot mode.*
- *Take your coffee or tea away from your desk and relish it.*

Laughter

Laughter brings pleasure. There is evidence that just an hour of loud, unrestrained laughter comprises "therapy" and improves symptoms of anxiety, stress, and depression.[9] Doing this regularly is an excellent "patch of pleasure." I would schedule at least half an hour of comedy viewing on television that forces you to laugh with abandon, every evening. A Japanese study examined the laughter habits of twenty thousand people aged over sixty-five. There was a clear relationship between the frequency of laughter and the prevalence of heart disease and stroke. Compared to those who laughed every day, 1.21 times as many people who hardly ever laughed suffered from heart disease. For stroke, the prevalence ratio was even higher, at 1.60.[10]

- *Laugh at every opportunity!*

Music

You can use music to fill your day with patches of pleasure. Pleasant music triggers the reward center.[11]

PREDICTIVE LISTENING

Our reward circuits buzz with pleasure when predictions we make from things we have observed turn out to be true. When you listen to a piece of music, you pick up its inherent rhythm. Once you identify it, you expect there to be a beat at the right points. Every time you are proved correct, you feel rewarded. If things become too predictable, you lose interest. A metronome, like a ticking clock, doesn't work beyond the first few moments for this reason. You need a little sparkle. A syncopation is when the inherent pattern of beats in a piece of music is momentarily interrupted by either a displaced or missing beat, forcing the listener to engage attention. Too many syncopations can cloud over the piece's inherent rhythm. With the right number of syncopations, the listener renews interest, forms a new prediction, and continues to enjoy the reward from testing the prediction at every beat. The "high" of the reward does not diminish over time.

Any situation that feels tedious or mundane may be helped along with music where the prediction of each beat sustains motivation and the syncopations attract attention. It may also evoke a feeling of being in control, which, as we've seen, lowers stress. Drum beat compositions, Indian tabla compositions, or jazz and rock rhythmic compositions would all serve this purpose.

> ■ Listen to drum beats with syncopations to keep you moti-
> vated as you work. Good choices include Japanese taiko drums,
> African drums such as the djembe, and the Indian tabla. Jazz
> and rock rhythms are also a good choice.

Hammering out a Rhythm

There is a group of artisans in Alappuzha, in Kerala, India, who create music while they build boats. Carpenters working on different sections of the boat synchronize their hammering and drilling as part of a giant percussion orchestra. Each hammer creates a unique note. Every tool plays a role in the orchestra. A carpenter synchronizes his rhythm with the underlying beat of the melody as he takes on a new task and every time a new task is begun, there is a syncopation for everyone. This setup provides ample pleasure and keeps everyone happy and stress-free.

THE ELVIS EFFECT

During his early days, Elvis Presley was famously threatened with arrest on account of his apparent inability to resist gyrating to his own music. This was deemed too risqué for his teenage audience. The situation became so bad that in 1956 arrest warrants were prepared ahead of a series of concerts in Jacksonville, Florida. Police lined the concert stage as Elvis performed and cameras watched his every move. Elvis later said that although he was bursting to move to the music, all he could do was wiggle his little finger. Wiggling that finger to the beat of his music brought him pleasure. He did not know it at the time, but this was a perfect demonstration of something we now know as "sensori-motor coupling," or the desperate desire we have to get into the "groove." Synchronizing our movements to a rhythm brings pleasure.

The biomechanics of every joint in the human body has an effect on the preferred frequency at which the body feels most comfortable moving. Walking feels best at a frequency of around 120 bpm or 2 Hz, which may be seen as the optimal walking pace.[12] This also happens to be the most perceived tempo in a broad spectrum of Western music.[13] The hip and shoulder joints also like to oscillate at this frequency. The elbows, knees, and fingers prefer to move faster. If we listen to a piece of music that has a tempo of within 3 to 15 percent of our preferred rhythm, it seems we "want" to synchronize some aspect of our movement to that rhythm. The greater our desire to move to music, the greater our pleasure when we do.

The greater our desire to move to music, the greater our pleasure when we do.

Haile Gebrselassie set a world record time of 4 minutes, 52.86 seconds while running the 2000m sprint in 1998. The "Elvis effect" may have played a role. Gebrselassie said afterward that he had synchronized his running pace to the rhythm in the song "Scatman," which was being played in the stadium in Birmingham, UK, right when he was running.

If you can't seem to find the motivation to go for a walk or a run, try playing rhythmic music. This will not only make you feel good, but the music will reduce your perceived exertion as you run. If you are exercising at low to moderate levels, music could make you feel up to 10 percent less tired compared to how you would feel without it. It may help you continue for longer by 15 percent. Studies have also shown that exercising to rhythmic music may improve efficiency so much that even oxygen consumption is reduced![14] According to an expert panel statement, a tempo of 125 to 140 bpm is optimal for motivating synchronized exercise in a young and healthy person.[15]

You may get an approximate idea of the pace of this range when you consider the hit song "From Paris to Berlin" by the band Infernal. When it is played on the radio, its tempo is 126 bpm, whereas the club version is played at 138 bpm. The popularity of a song like this may arise in part from its tempo range, which resonates with the frequency of human movement.

- *Walk, jog, run, or dance to music whenever you can. If you are in a gym, watch yourself in a mirror as you move, noting your synchrony with your music.*
- *If you need motivation to go for a run, choose a tune with empowering lyrics that has a frequency of 125 to 140 bpm.*

Passing the Time

Faster, arousing music makes time pass more slowly.[16] One theory behind this is that we have an internal clock that keeps time. Faster music makes it "tick" faster, so it overestimates the passage of time.[17] You might feel you have been waiting for an hour when you have only been waiting for fifteen minutes. On the other hand, slow music makes us underestimate time, and an hour might fly by in what seems like fifteen minutes. To make time pass quickly, listen to music that is relaxing and slow.[18]

Gratitude

When you appreciate something, you are forcing yourself to see it in a positive light. As you search for a reason to feel gratitude, you are switching your attentional focus from negative sensations to positive ones and you feel pleasure. You can create a moment of pleasure by devoting fifteen minutes of your evening to appreciating the good in something, be it a person or an event. It can be something as simple as someone holding the elevator doors for you as you come racing down the hall, or it can be the security guard in your office building paying you a compliment. Dwell on the positive emotions you experienced during your interaction with the person or event. Spending fifteen minutes every night dwelling on something positive that has happened during the day forms part of a technique for treating anhedonia in patients with schizophrenia.[19]

- *Spend fifteen minutes every night describing and dwelling on one positive thing you experienced that day.*

⊙ PLEASURE AND PAIN

In this chapter I've tried to convince you of the importance of pleasure but remember, there can be too much of a good thing.

When the stresses of life grow unbearable, some of us may

feel drawn toward finding instant and powerful sources of pleasure. We tell ourselves we deserve these "pleasure kicks" as a justified reward for making it through the day or week. They catapult us from our grey world into a world of dazzling color. They remind us what euphoric happiness feels like. We are so moved by the experience that we are desperate to experience it again, just one last time, *every time.*

If chronic stress changes the brain such that pleasure becomes more difficult to experience, it is inevitable that we resort to novel and more potent sources of pleasure that will fish us out of a vortex of gloom. This is how drug and alcohol abuse begins. Recreational drugs (such as cocaine and amphetamine) and alcohol powerfully activate the reward circuits. The potency of this effect draws the user back time and time again. The problem is, the reward circuit's wiring and signals start to change and you require more of the drug to feel the same. Eventually even your baseline of normalcy feels more unpleasant than it did *before* you started using the drug. You now reach for the drug to just make your baseline feel "normal" again. While at first you resorted to these substances in order to feel joy, you now rely on them in order to not feel sad.[20][21] The neural circuitry of addiction is complex and involves many networks and messengers. Some of these overlap with your stress circuitry and addiction can worsen the effects of chronic stress.

■ *Choose your sources of pleasure wisely or your quest for pleasure will turn into a quest to escape pain.*

CHAPTER 9

Aligning Your Beliefs and Goals for Long-Term Success

ABOUT TWO DECADES ago, the world of brain science uncovered an interesting paradox. You would expect that the brain of a person with his feet up, thinking of nothing and doing nothing, would work less hard than the cerebral matter of a mathematician trying to solve an impossible equation. But that is not what Gordon Shulman, a professor of neurology at Washington School of Medicine, discovered.[1] He found that when a person relaxes, instead of the entire brain becoming quieter, a distinct network in the brain becomes *more* active. His team wrote a landmark paper in 2001 in which they labeled this peculiar network's activity (it becomes "busy" when the brain does "nothing" and "falls asleep" when the brain "works hard") the brain's "default mode."[2] We now call the network the default mode network, or DMN.

Your DMN lights up under a brain scan when you are thinking about anything related to yourself. The network is spread along the midline of the brain. It works together with networks in regions such as the hippocampus to create an autobiographical representation by weaving together your memories, ongoing experiences, and perceived sensations into a story in which you are the lead character. Your autobiographical memory gives form to the concept of *you*. Your DMN is analogous to what Sigmund Freud referred to as the *ego*.[3] Since your experiences shape who you are and what you believe in, your core beliefs and long-term vision of yourself are likely to be shaped by your autobiographical memory and by your DMN. Your autobiographical memory contributes to your sense of "self."

Your DMN's musings are where your mind heads over to, whenever it wanders. Your DMN is always planning, imagining, and projecting. It is engaged in prospection and retrospection as it uses your past to predict your future. Before every occasion of uncertainty, your DMN carries out a series of simulations so you are prepared for every possibility. As soon as something in your external world engages you, you disengage from your DMN. If you are bored and your mind wanders, you reengage with it. If you are good at regulating your emotions and don't give negative emotions much airtime, and if you make sure you experience pleasure regularly, your autobiographical memory will be filled with happy stories. If not, it will likely have a negative bias. If it has a negative bias *and* you are prone to mind-wandering, you will be more likely to dwell on negative thoughts and make negative projections every time your mind wanders, which will add to your stress burden.

There are three broad areas that are influenced by your DMN and that you may be able to manipulate to reduce your stress burden. These are:

- Your core beliefs
- Your long-term goals
- Your sense of self

In contributing to your autobiographical memory, your DMN shapes your core beliefs and likes and dislikes. These ripple along either in or out of phase with the life you lead. Some psychoanalytic theorists such as Sigmund Freud and Carl Jung proposed that when you suppress your innermost longings and beliefs for the sake of living the life you are expected to lead, depression and other mental illnesses surface.

If your life or work contradicts your core beliefs, you feel unhappy and this increases stress. In his TEDx talk "First Why and Then Trust," the leadership expert Simon Sinek elucidates how an organization that is starting out thrives while its core beliefs align with the core beliefs of its employees. As it grows too large, these beliefs are obfuscated and it is at that point (which Sinek calls "the split") when employees start suffering from stress and unhappiness.

© COGNITIVE DISSONANCE

You might experience *cognitive dissonance* if you come across new, verified information that contradicts a core belief or idea that you hold and that forces you to hold two or more conflicting ideas at the same time. Cognitive dissonance arises out of inconsistency, when what you think, say, and do, don't match. Cognitive dissonance contributes to your stress burden.

Happiness is when what you think, what you say and what you do are in perfect harmony.   —M. K. Gandhi

For instance, you might think you ought to spend time at home with your young child, but you also think that putting the hours in at work will accelerate your career. While you are at work, you think you should be with your child and when you are with your child, you feel you should be thinking about your work. Your core belief in honesty might be challenged when you are employed by a company with dubious ethical practices. You muffle your inner conflict as you work with your team to churn out profits made on dishonest grounds. Here, what you are thinking does not match what you are doing and saying in public. There are several ways to reduce cognitive dissonance.

- You can form a third idea by combining two or more conflicting ideas.
- You can change your core beliefs to bring them in line with your life.
- You can change your life to bring it in line with your core beliefs.

If you are forced to hold two conflicting ideas, you need a third idea to reconcile the two. When peace-loving human beings must kill during times of war, they face cognitive dissonance. Viewing world peace as the ultimate goal that cannot be achieved without bloodshed might bring the two conflicting views in line. Many working mothers reduce the cognitive dissonance that arises from thinking they should be at work *and* at home with their child by adopting a unifying higher goal that reconciles both ideas, such as seeing their work as their contribution to improve gender equality in the world their children will inherit. Directing the focus away from yourself and toward others—especially toward *serving others*—and bringing about *global good* can reduce the burden of stress.

If you can't change your life and it grates against your core beliefs, you can try changing your core beliefs. The phenomenon of "sour grapes" is an example of this. If you can't change your core beliefs, you have to either change your life or incorporate realigning actions that bring your life in line with your beliefs. As an example, if you have done something that *you know* is wrong but you can't admit it to the world, "confessing" to it (in private) brings what you say and do in line with what you think, if only momentarily. This may explain why journal writing can offer respite from mental distress. Speaking your mind can release cognitive dissonance because what you say and do echoes your thoughts. Sometimes, a cathartic quarrel venting pent-up resentment can make you feel better because it rids you of the dissonance caused by hiding what you feel.

Music That Soothes the Conflicted Mind

Mozart's compositions are said to relieve cognitive dissonance, perhaps because we learn how to reconcile dissonance when conflict is presented to us in musical form and the composer leads us to its resolution. Actively listening to Mozart's compositions (or to similar classical music) for half an hour every evening, or whenever you can, might reduce your cognitive dissonance load. I would try to follow the music carefully, as if you were watching a live performance, and try to lose yourself for that half hour. Music played in a major key sounds happy; a minor key sounds sad.

- Look out for cognitive dissonance, and resolve it.
- Spend some time listening to the compositions of Mozart, every day.

☺ LONG-TERM GOALS

A good writer visualizes the story line ahead of what he is writing. It helps to know where the lead character is heading. This gives the character purpose. As your brain manufactures your autobiographical memory, it carries out *autobiographical planning*— it plans ahead.

> *He who has a Why to live for can bear almost any How.*
> —*Friedrich Nietzsche*

When you go about your life, you operate on two dimensions. One dimension relates to your efforts to meet immediate or short-term goals. A second dimension operates on a higher plane. This dimension involves your overall view of life, your vision, and your long-term goals. A long-term goal is reached with many small steps. Each of those steps is a short-term goal. So, you operate in a short-term dimension while you navigate along your long-term dimension.

There is some evidence that if your long-term goals are potent enough, you can "rise above" short-term difficulties and setbacks. If you envision having to endure a difficult colleague for the next three months for the sake of getting the training you need for your long-term goal, you are more likely to pacify your distress than if you have no goal and must face the colleague every day without any good coming out of it.

Having a sense of purpose correlates with general good health.

Having a sense of purpose correlates with general good health, including lower levels of the inflammatory agent IL-6 and a reduced risk of Alzheimer's disease, heart attacks, and stroke. Men and women who survive through exceptionally trying times often describe holding on to a long-term vision or purpose that overrode short-term distress and carried them through. Viktor Frankl, an eminent psychiatrist, neurologist, and Holocaust survivor, described this idea when he wrote:

"The experiences of camp life show that man does have a choice of action . . . become a plaything of circumstances . . . or strive and struggle for a worthwhile goal . . ."[4]

The separation of the short-term dimension from the long-term dimension is also underscored by the words of a former prisoner of war from the Vietnam War, James Stockdale.[5] In an interview, he was asked who did not survive Vietnam. His heart-breaking reply was "Oh, that's easy, the optimists. Oh, they were the ones who said, 'We're going to be out by Christmas.' And Christmas would come, and Christmas would go. Then they'd say, 'We're going to be out by Easter.' And Easter would come, and Easter would go. And then Thanksgiving, and then it would be Christmas again. And they died of a broken heart."

Stockdale's humbling account suggests that shifting focus to the longer-term picture might have helped people bear repeated shorter-term disappointments a little better.

- Focus on the big picture.
- Visualize and prioritize a long-term goal.
- Detach from your short-term dimension and immerse yourself in the long-term dimension, whenever possible.
- If you reach one long-term goal, immediately seek a new one.

SELF-EFFICACY

Self-efficacy is the belief that you have the capability to succeed and reach your goals.[6] It is negatively correlated with post-traumatic stress and general psychological distress.[7] Belief in your own ability is born partly out of your autobiographical memory, so you can build self-efficacy by filling your autobiography with challenges that you have managed to overcome.

These can take any form. Sporting challenges can increase your sense of achievement and signing up for demanding physical challenges will complement rather than interfere with a stressful occupation.

- *Regularly replenish your collection of successfully completed challenges. Sports provide an easy and accessible option.*

◎ GROWTH MINDSET

There are two ways of thinking about your "self":

1. You cannot change. If you can't do something, you will never be able to do it.
2. Like a small child, you can "grow" to do anything. Anything you can't do is simply a consequence of not having learned to do it *yet*.

Carol Dweck, a professor of psychology at Stanford University whose extensive work on theories of intelligence has challenged traditional views about education, personal achievement, and success, describes these two mindsets as a "fixed mindset" and a "growth mindset." In her book *Mindset: The New Psychology of Success*, she describes how switching from one mindset to another can drastically change a person's life. Dr. Dweck cites the examples of some of the greatest athletes in history who, contrary to expectations, were not *born* talented but *became* so. When facing rival competitors with greater talent, their mental determination and self-belief charged them to victory.

If you hold a fixed mindset, you will view yourself as static rather than as being in a state of constant improvement. A weakness will seem like an ineffable flaw and not an area you can work on and improve. You may feel inclined to hide your weaknesses, and situations where they are at risk of surfacing will cause you unnecessary stress. If you hold a growth mindset, you will see a weakness as a temporary state through which you are passing, so it does not reflect your "worthiness." Situations where your weaknesses may be exposed will stress you far less. If you encounter an unwanted outcome, adopting a growth mindset will reduce your self-blame and self-pity. If you fail in a test, you won't take the result of that test as an immutable proclamation of your worth. If you acknowledge you are in a state of constant growth, your stress response will be reduced in competitive or socio-evaluative situations, where you feel you are being "judged."

- *Adopt a growth mindset.*
- *Identify instances when you lapse into a fixed mindset.*

◎ SELF-OPINION

Given how much time you spend mulling over your autobiography every time your mind wanders, it is important to *like* yourself. Negative thoughts generate further negative thoughts and push you down a rabbit hole of negative emotions, activating your emotional brain. If you do things that make you privately see yourself as unkind or as a "bad" person, you won't like the autobiography that is being written. Conversely, if you stock up on "good stories" whenever you can, your autobiography will glow with self-esteem. No matter what others say, you *know* you are a good person.

Dr. J. Robert Oppenheimer was a broken man when he

discovered he had been successful at creating the world's first atomic bomb. He clutched at all the good he had ever done, to remind himself that although he was now about to be implicated in death and destruction, albeit indirectly, he was primarily a "good person." Prior to the Trinity test, months before the tragedies of Hiroshima and Nagasaki, he said the following words:

> In battle, in the forest, at the precipice in the mountains,
> On the dark great sea, in the midst of javelins and arrows,
> In sleep, in confusion, in the depths of shame,
> The good deeds a man has done before, defend him.
>   —Dr. J. Robert Oppenheimer
>   (quoting from the Bhagavad Gita)

- *Do good, as often as you can.*
- *If you don't like yourself, change.*

ⓒ SELF-COMPASSION

Viewing yourself as a vulnerable person or "child" has shown promise in reducing depressive symptoms.[8] In one experiment, some volunteers with major depressive disorder were put into a despondent mood in a laboratory from reading some depressing self-referential statements such as "I think I am a loser," listening to melancholic music such as Tomaso Albinoni's "Adagio for Strings and Organ in G minor," and being told to dwell on negative thoughts. When they were treated with cognitive therapy afterward, the therapy worked better if they had filled their minds with thoughts of self-compassion beforehand. Part of the instruction the authors of the study gave to the volunteers was,

"Try to see yourself from an outsider's point of view, from the perspective of a compassionate, friendly observer."

■ *Visualize a scene in which you step out of your skin and approach yourself as a kind and compassionate friend.*
■ *Dwell on the scene as you show compassion to yourself.*

A Final Note on Resilience

Resilience is the art of *bouncing back*. The secret of navigating our complex world—rocked with setbacks, shaken with trauma, and exhausted with the daily grind of work—lies in a single concept: *elasticity*. All rubber bands have an elastic limit. They keep springing back unless you stretch them beyond this limit. This book has shown you how to increase your elastic limit, so you can keep bouncing back.

History affords us an anthology of remarkable resilience in people from diverse cultures and different times. There seem to be patterns and traits that weave through each story. In 2003, Dr. Kathryn Connor and Dr. Jonathan Davidson of Duke University studied many such anecdotes and created what is known as the Connor–Davidson Resilience Scale. Based on this scale, some authors have proposed targeting the following areas with psychotherapeutic techniques to nurture some of the psychosocial aspects of resilience:[1]

- Optimism
- Interpretation

- Self-opinion
- Physical fitness
- Active coping
- A social network

◎ OPTIMISM

Resilience is fostered when we're optimistic about the overall picture or about the eventual future.

Setbacks that you experience here and now are offset by optimism for the future. People with **grit** have an unshakable long-term aspiration or hope that keeps them going in the face of countless failures and disappointments miring the present.

◎ INTERPRETATION

Cognitive reappraisal reduces the emotional distress a situation presents. Seeing an event from the perspective of your actions rather than as a reflection of you, and viewing a setback as a signal for the need for growth, rather than as a declaration of failure, tricks your mind into viewing the event positively.

◎ SELF-OPINION

It's important to believe that you are a good person and that you have a sound moral compass. Your self-opinion should be garnished with memories of how strong you have been, and hence can be, in a crisis situation.

◎ PHYSICAL FITNESS

Exercise confers a kind of psychological cross-protection. If you are confident physically, you are more likely to feel mentally confident during a challenge.

◎ ACTIVE COPING

When you meet a setback or crisis, it's essential to respond by taking action in some way, instead of lapsing into a state of emotional inertia. Active, rather than passive, coping confers resilience.

◎ A SOCIAL NETWORK

A social scaffold offers safety and acceptance and may reinforce self-belief. It may also present opportunities for meeting resilient individuals who can act as role models. Post-traumatic stress disorder in Iraq and Afghanistan war veterans is often associated with less social support, broken relationships, and the absence of a strong social network.

◎ A LITTLE WILL GO A LONG WAY

Human existence tells a story of serial upgrades as we become better as a race and better as individuals at making the most of the life we are born into. As that life changes from one generation to the next, we change too. This ability to adapt is a sign of human intelligence and is conferred by the plastic nature of our brains.

Technological progress, changes in societal structure, and globalization are rapidly altering the way we live. If change occurs faster than the pace of our adaptation, we adapt badly and may maladapt to change. This maladaptation leaves dents in a broad array of systems, from brain circuitry and hormonal dynamics to metabolism and immune regulation.

Although resilience is best achieved with team effort, by addressing all of these systems simultaneously, you may not always be able to implement all the suggestions described in this book at once. For instance, if you are an airline pilot, your circadian regularity may be out of your hands; if you are a soldier in active service, it may be impossible to keep your sympathetic tone low; if you work in palliative care, it may be challenging to practice emotional regulation; if you are a long-distance truck driver, you may not be able to choose what or when you eat.

Controlling what you *can* control will buffer the toll taken by what you *can't* control. The advice in this book is there to serve you, rather than enslave you. Whatever little you do will add to your resilience reserve. If you can't do everything, do what you can. It will go a long way. Good luck!

Acknowledgments

This book would not have been written without the unconditional love of my husband, Laurent.

I cannot thank Andrea Somberg enough for her extraordinary diligence, optimism, and energy. Despite being on the opposite side of the globe, Andrea has been only moments away throughout this project, at all hours of the day, every day of the year, tirelessly offering advice and assistance whenever needed.

A big thank you to Marian Lizzi for her immense patience, dedication, and care in editing the manuscript. Thank you also to Lauren Appleton, Ian Gibbs, and the entire team at Tarcher-Perigee for turning the manuscript into a beautiful book.

I thank Francois Rigou for his enthusiasm and encouragement when this book was still an idea, and for test-reading an early draft of the manuscript. Thank you to Kelly Tagore for her warm support and brilliant advice.

I am very grateful to Craig Goldsmith, Rishaad Salamat, and Alex Storoni for providing me with valuable feedback on the manuscript. Thank you also to Greg Siegle for molding my knowledge of the prefrontal cortex with his wisdom, and to Gordon Plant for the continued privilege of learning from his genius.

I am very grateful for the cups of tea and encouragement offered by all my wonderful friends, especially Pat Barech, Tali and Dan Ezra, Annie Rigou, Margaret-Mary O'Brien, Matthieu Robert, Harijs Deksnis, Andrew Edwards, and Alli McCoy.

Finally, I would like to thank Paul Hegarty, whose remarkable strength in the face of grave adversity convinced me that resilience can be cultivated to endure the most trying of challenges.

Notes

INTRODUCTION

1   http://www.guinnessworldrecords.com/world-records/fastest-half
-marathon-barefoot-on-icesnow.
2   M. Kox, L. T. van Eijk, J. Zwaag, J. van den Wildenberg, F. C. Sweep,
J. G. van der Hoeven, and P. Pickkers, "Voluntary activation of the
sympathetic nervous system and attenuation of the innate immune
response in humans," *Proceedings of the National Academy of Sciences of the United States of America* 111, no. 20 (May 2014): 7379–84.
3   The Canadian Medical Hall of Fame http://cdnmedhall.org/inductees
/dr-hans-selye.
4   H. Selye, *The Stress of Life* (New York: McGraw-Hill, 1956).
5   B. S. McEwen, "Stressed or Stressed Out: What is the Difference?"
Journal of Psychiatry and Neuroscience 30, no. 5 (2005): 315–18.
6   Peter Sterling, "Principles of Allostasis: Optimal Design, Predictive
Regulation, Pathophysiology and Rational Therapeutics," in J. Schulkin,
ed., *Allostasis, Homeostasis, and the Costs of Adaptation*, (Cambridge University Press, 2004).

CHAPTER 1

1   A. Etkin, T. Egner, D. M. Peraza, E. R. Kandel, and J. Hirsch, "Resolving emotional conflict: a role for the rostral anterior cingulate

cortex in modulating activity in the amygdala," *Neuron* 51, no. 6 (Sept. 2006): 871–82.

2	A. Golkar, E. Johansson, M. Kasahara, W. Osika, A. Perski, and I. Savic, "The Influence of Work-related Chronic Stress on the Regulation of Emotion and on Functional Connectivity in the Brain," *PLoS ONE* 9, no. 9 (Sept. 2014): e104550.

3	N. Sadeh, J. M. Spielberg, M. W. Miller, W. P. Milberg, D. H. Salat, M. M. Amick, C.B. Fortier, and R. E. McGlinchey. "Neurobiological indicators of disinhibition in posttraumatic stress disorder," *Human Brain Mapping* 36, no. 8 (Aug. 2015): 3076–86.

4	F. Beissner, K. Meissner, K. J. Bär, and V. Napadow, "The autonomic brain: an activation likelihood estimation meta-analysis for central processing of autonomic function," *Journal of Neuroscience* 33, no. 25 (Jun. 2013): 10503–11.

5	V. G. Macefield, C. James, and L. A. Henderson, "Identification of sites of sympathetic outflow at rest and during emotional arousal: concurrent recordings of sympathetic nerve activity and fMRI of the brain," *International Journal of Psychophysiology* 89, no. 3 (Sept. 2013): 451–9.

6	A. F. Arnsten, "Stress Weakens Prefrontal Networks: Molecular Insults to Higher Cognition," *Nature Neuroscience* 18, no. 10 (Oct. 2015): 1376–85, doi: 10.1038/nn.4087.

7	I. Negrón-Oyarzo, F. Aboitiz, and P. Fuentealba, "Impaired Functional Connectivity in the Prefrontal Cortex: A Mechanism for Chronic Stress-induced Neuropsychiatric Disorders," *Neural Plasticity* 2016 (2016): Article ID 7539065.

8	J. J. Radley, R. M. Anderson, B. A. Hamilton, J. A. Alcock, and S. A. Romig-Martin, "Chronic Stress-induced Alterations of Dendritic Spine Subtypes Predict Functional Decrements in an Hypothalamo-pituitary-adrenal-inhibitory Prefrontal Circuit," *Journal of Neuroscience* 33, no. 36 (Sept. 2013): 14379–91.

9	Y. C. Tse, I. Montoya, A. S. Wong, A. Mathieu, J. Lissemore, D. C. Lagace, and T. P. Wong, "A Longitudinal Study of Stress-induced Hippocampal Volume Changes in Mice That Are Susceptible or Resilient to Chronic Social Defeat," *Hippocampus* 24, no. 9 (Sept. 2014): 1120–8.

10	A. Starčević, I. Dimitrijević, M. Aksić, L. Stijak, V. Radonjić, D. Aleksić, and B. Filipović, "Brain Changes in Patients with Posttraumatic Stress Disorder and Associated Alcoholism: MRI Based Study," *Psychiatria Danubina* 27, no. 1 (Mar. 2015): 78–83.

11 L. H. Rubin, V. J. Meyer, R. J. Conant, E. E. Sundermann, M. Wu, K. M. Weber, M. H. Cohen, D. M. Little, and P. M. Maki, "Prefrontal Cortical Volume Loss is Associated with Stress-related Deficits in Verbal Learning and Memory in HIV-infected Women," *Neurobiology of Disease* (Sept. 2015), pii: S0969-9961(15)30056-5.

12 A. Vyas, R. Mitra, B. S. Shankaranarayana Rao, and S. Chattarji, "Chronic Stress Induces Contrasting Patterns of Dendritic Remodeling in Hippocampal and Amygdaloid Neurons," *Journal of Neuroscience* 22 (2002): 6810–18.

13 G. L. Moreno, J. Bruss, and N. L. Denburg, "Increased Perceived Stress is Related to Decreased Prefrontal Cortex Volumes among Older Adults," *Journal of Clinical and Experimental Neuropsychology* (Sept. 2016): 1–13.

14 W. C. Drevets, "Neuroimaging and Neuropathological Studies of Depression: Implications for the Cognitive-Emotional Features of Mood Disorders," *Current Opinion in Neurobiology* 11 (2001): 240–49.

15 G. Seravalle and G. Grassi, "Sympathetic Nervous System, Hypertension, Obesity and Metabolic Syndrome," *High Blood Pressure & Cardiovascular Prevention* 23, no. 3 (Sept. 2016): 175–9.

16 B. M. Egan, "Insulin Resistance and the Sympathetic Nervous System," *Current Hypertension Reports* 5, no. 3 (Jun. 2003): 247–54.

17 V. Zotev, R. Phillips, K. D. Young, W. C. Drevets, and J. Bodurka, "Prefrontal Control of the Amygdala During Real-time fMRI Neurofeedback Training of Emotion Regulation." *PLoS ONE* 8 (2013): e79184.

18 E. Fuchs, G. Flugge, and B. Czeh, "Remodeling of Neuronal Networks by Stress." *Frontiers in Bioscience* 1, no. 11 (Sept. 2006): 2746–58.

19 R. S. Duman, "Pathophysiology of Depression and Innovative Treatments: Remodeling Glutamatergic Synaptic Connections." *Dialogues in Clinical Neuroscience* 16, no. 1 (Mar. 2014): 11–27.

20 S. L. Christiansen, K. Højgaard, O. Wiborg, and E. V. Bouzinova EV, "Disturbed Diurnal Rhythm of Three Classical Phase Markers in the Chronic Mild Stress Rat Model of Depression." *Neuroscience Research* 110 (Sept. 2016): 43–8.

21 Y. Wu, L. Dissing-Olesen, B. A. MacVicar, and B. Stevens, "Microglia: Dynamic Mediators of Synapse Development and Plasticity," *Trends in Immunology* 36, no. 10 (2015): 605–13.

22 A. Kleinridders, H. A. Ferris, W. Cai, and C. R. Kahn, "Insulin Action in Brain Regulates Systemic Metabolism and Brain Function," *Diabetes* 63, no. 7 (Jul. 2014): 2232–43.

23 A. J. Loonen and S. A. Ivanova, "Circuits Regulating Pleasure and Happiness-Mechanisms of Depression," *Frontiers in Human Neuroscience* 10, no. 10 (Nov. 2016): 571.

CHAPTER 2

1 Brian M. Galla and Jeffrey J. Wood, "Trait Self-Control Predicts Adolescents' Exposure and Reactivity to Daily Stressful Events," *Journal of Personality* 83, no. 1 (Feb. 2015): 69–83.

2 T. D. Wager, M. L. Davidson, B. L. Hughes, M. A. Lindquist, and K. N. Ochsner, "Prefrontal-subcortical Pathways Mediating Successful Emotional Regulation," *Neuron* 59 (2008): 1037–50.

3 E. Blix, A. Perski, H. Berglund, and I. Savic, "Long-term Occupational Stress is Associated with Regional Reductions in Brain Tissue Volumes," *PLoS ONE* 8 (2013): e64065.

4 M. Koenigs, E. D. Huey, M. Calamia, V. Raymont, D. Tranel, and J. Grafman, "Distinct Regions of Prefrontal Cortex Mediate Resistance and Vulnerability to Depression," *Journal of Neuroscience* 28 (2008): 12341–48.

5 T. S. Ligeza, M. Wyczesany, A. D. Tymorek, and M. Kamiński, "Interactions between the Prefrontal Cortex and Attentional Systems during Volitional Affective Regulation: An Effective Connectivity Reappraisal Study," *Brain Topography* 29, no. 2 (Mar. 2016): 253–61.

6 G. Sheppes and Z. Levin, "Emotion Regulation Choice: Selecting between Cognitive Regulation Strategies to Control Emotion," *Frontiers in Human Neuroscience* 7 (2013): 179.

7 R. B. Price, B. Paul, W. Schneider, and G. J. Siegle, "Neural Correlates of Three Neurocognitive Intervention Strategies: A Preliminary Step Towards Personalized Treatment for Psychological Disorders," *Cognitive Therapy and Research* 37, no. 4 (2013): 657–72.

8 M. Csíkszentmihályi and J. LeFevre, "Optimal Experience in Work and Leisure," *Journal of Personality and Social Psychology* 56, no. 5 (May 1989): 815–22.

9 A. Manna, A. Raffone, M. G. Perrucci, D. Nardo, A. Ferretti, A. Tartaro, A. Londei, C. Del Gratta, M. O. Belardinelli, and G. L. Romani, "Neural Correlates of Focused Attention and Cognitive Monitoring in Meditation," *Brain Research Bulletin* 82, nos. 1–2 (Apr. 2010): 46–56.

10 R. Nouchi, Y. Taki, H. Takeuchi, H. Hashizume, T. Nozawa, T. Kambara, A. Sekiguchi, C. M. Miyauchi, Y. Kotozaki, H. Nouchi, and R. Kawashima, "Brain Training Game Boosts Executive Functions,

Working Memory and Processing Speed in the Young Adults: A Randomized Controlled Trial," *PLoS ONE* 8, no. 2 (2013): e55518.

11 M. Muraven, R. F. Baumeister, and D. M. Tice, "Longitudinal Improvement of Self-regulation through Practise: Building Self-control through Repeated Exercise," *Journal of Social Psychology* 139 (1999): 446–57.

12 M. Muraven, "Practicing Self-control Lowers the Risk of Smoking Lapse," *Psychology of Addictive Behaviors* 24 (2010): 446–52.

13 M. Sakaki, H. J. Yoo, L. Nga, T. H. Lee, J. F. Thayer, and M. Mather, "Heart Rate Variability is Associated with Amygdala Functional Connectivity with MPFC across Younger and Older Adults," *Neuroimage* 31, no.139 (May 2016): 44–52.

14 J. F. Thayer, A. L. Hansen, E. Saus-Rose, and B. H. Johnsen, "Heart Rate Variability, Prefrontal Neural Function, and Cognitive Performance: The Neurovisceral Integration Perspective on Self-regulation, Adaptation, and Health," *Annals of Behavioral Medicine* 37, no. 2 (Apr. 2009): 141–53.

15 S. C. Segerstrom and L. S. Nes, "Heart Rate Variability Reflects Self-regulatory Strength, Effort, and Fatigue," *Psychological Science* 18, no. 3 (Mar. 2007): 275–81.

16 T. F. Heatherton and D. D. Wagner, "Cognitive Neuroscience of Self-regulation Failure," *Trends in Cognitive Sciences* 15, no. 3 (Mar. 2011): 132–39.

17 P. Bermudez et al., "Neuroanatomical Correlates of Musicianship as Revealed by Cortical Thickness and Voxel-based Morphometry," *Cerebral Cortex* 19 (2009): 1583–96.

18 K. Houben, F. C. Dassen, and A. Jansen, "Taking Control: Working Memory Training in Overweight Individuals Increases Self-regulation of Food Intake," *Appetite* 105 (Oct. 2016): 567–74.

19 J. Cranwell, S. Benford, R. J. Houghton, M. Golembewksi, J. E. Fischer, and M. S. Hagger, "Increasing Self-Regulatory Energy Using an Internet-based Training Application Delivered by Smartphone Technology," *Cyberpsychology, Behavior and Social Networking* 17, no. 3 (2014): 181–86.

20 M. H. Ashcraft and E. P. Kirk, "The Relationships among Working Memory, Math Anxiety, and Performance," *Journal of Experimental Psychology: General* 130 (2001): 224–37.

21 C. B. Young, S. S. Wu, and V. Menon, "The Neurodevelopmental Basis of Math Anxiety," *Psychological Science* 23 (2012): 492–501.

22 I. M. Lyons and S. L. Beilock, "When Math Hurts: Math Anxiety Predicts Pain Network Activation in Anticipation of Doing Math," *PLoS ONE* 7 (2012a): e48076.

23 A. Sarkar, A. Dowker, and R. Cohen Kadosh, "Cognitive Enhancement or Cognitive Cost: Trait-Specific Outcomes of Brain Stimulation in the Case of Mathematics Anxiety," *Journal of Neuroscience* 34.50 (2014): 16605–10.

24 B. J. Casey, L. H. Somerville, I. H. Gotlib, O. Ayduk, N. T. Franklin, M. K. Askren, J. Jonides, M. G. Berman, N. L. Wilson, T. Teslovich, G. Glover, V. Zayas, W. Mischel, and Y. Shoda, "Behavioral and Neural Correlates of Delay of Gratification 40 Years Later," *Proceedings of the National Academy of Sciences of the United States of America* 108, no. 36 (Sept. 2011): 14998–15003.

25 T. P. Alloway and J. C. Horton, "Does Working Memory Mediate the Link Between Dispositional Optimism and Depressive Symptoms?" *Applied Cognitive Psychology* 30, no. 6 (Nov./Dec. 2016): 1068–72.

26 A. Curci, T. Lanciano, E. Soleti, and B. Rimé, "Negative Emotional Experiences Arouse Rumination and Affect Working Memory Capacity," *Emotion* 13, no. 5 (Oct. 2013): 867–80.

27 L. Xiu, R. Zhou, and Y. Jiang, "Working Memory Training Improves Emotion Regulation Ability: Evidence from HRV," *Physiology & Behavior* 155 (Dec. 2015): 25–29.

28 S. Kühn, T. Gleich, R. C. Lorenz, U. Lindenberger, and J. Gallinat, "Playing Super Mario Induces Structural Brain Plasticity: Gray Matter Changes Resulting from Training with a Commercial Video Game," *Molecular Psychiatry* 19, no. 2 (Feb. 2014): 265–71.

29 T. S. Ligeza, M. Wyczesany, A. D. Tymorek, and M. Kamiński, "Interactions between the Prefrontal Cortex and Attentional Systems during Volitional Affective Regulation: An Effective Connectivity Reappraisal Study," *Brain Topography* 29, no. 2 (Mar. 2016): 253–61.

30 C. A. Ray, and K. M. Hume, "Neck Afferents and Muscle Sympathetic Activity in Humans: Implications for the Vestibulosympathetic Reflex," *Journal of Applied Physiology* (1985) 84, no. 2 (Feb. 1998): 450–53.

31 J. C. Geinas, K. R. Marsden, Y. C. Tzeng, J. D. Smirl, K. J. Smith, C. K. Willie, N. C. Lewis, G. Binsted, D. M. Bailey, A. Bakker, T. A. Day, and P. N. Ainslie, "Influence of Posture on the Regulation of Cerebral Perfusion," *Aviation, Space, and Environmental Medicine* 83, no. 8 (Aug. 2012): 751–57.

32 L. A. Uebelacker, G. Epstein-Lubow, B. A. Gaudiano, G. Tremont, C. L. Battle, and I. W. Miller, "Hatha Yoga for Depression: Critical Review of the Evidence for Efficacy, Plausible Mechanisms of Action, and

Directions for Future Research," *Journal of Psychiatric Practice* 16, no. 1 (Jan. 2010): 22–33.

33 Alain, "Propos sur le bonheur," *Gallimard*, Folio Essais 21 (1928): 11–13.

34 J. A. Robinson and K. L. Swanson, "Field and Observer Modes of Remembering," *Memory* 1, no. 3 (Sept. 1993): 169–84.

35 K. E. Gilbert, S. Nolen-Hoeksema, and J. Gruber, "Positive Emotion Dysregulation across Mood Disorders: How Amplifying versus Dampening Predicts Emotional Reactivity and Illness Course," *Behaviour Research and Therapy* 51, no. 11 (Nov. 2013): 736–41.

36 E. Watkins, "Adaptive and Maladaptive Ruminative Self-focus during Emotional Processing," *Behaviour Research and Therapy* 42, no. 9 (Sept. 2004): 1037–52.

37 J. van Lier, M. L. Moulds, and F. Raes, "Abstract 'Why' Thoughts about Success Lead to Greater Positive Generalization in Sport Participants," *Frontiers in Psychology* 6 (Nov. 2015): 1783.

38 F. Wang, C. Wang, Q. Yin, K. Wang, D. Li, M. Mao, C. Zhu, and Y. Huang, "Reappraisal Writing Relieves Social Anxiety and May be Accompanied by Changes in Frontal Alpha Asymmetry," *Frontiers in Psychology* 21, no. 6 (Oct. 2015): 1604.

39 R. F. Helfrich and R. T. Knight, "Oscillatory Dynamics of Prefrontal Cognitive Control," *Trends in Cognitive Science* 20, no. 12 (Dec. 2016): 916–30.

40 K. Song, M. Meng, L. Chen, K. Zhou, and H. Luo, "Behavioral Oscillations in Attention: Rhythmic α Pulses Mediated Through θ Band," *Journal of Neuroscience* 34, no.14 (Apr. 2014): 4837–44.

41 H. S. Lee, A. Ghetti, A. Pinto-Duarte, X. Wang, G. Dziewczapolski, F. Galimi, S. Huitron-Resendiz, J. C. Piña-Crespo, A. J. Roberts, I. M. Verma, T. J. Sejnowski, and S. F. Heinemann, "Astrocytes Contribute to Gamma Oscillations and Recognition Memory," *Proceedings of the National Academy of Sciences of the United States of America* 111, no. 32 (Aug. 2014): E3343–52.

42 P. Billeke, F. Zamorano, D. Cosmelli, and F. Aboitiz, "Oscillatory Brain Activity Correlates with Risk Perception and Predicts Social Decisions," *Cerebral Cortex* 23, no. 12 (Dec. 2013): 2872–780.

43 B. Voloha, T. Valiante, S. Everling, and T. Womelsdorf, "Theta-gamma Coordination between Anterior Cingulate and Prefrontal Cortex Indexes Correct Attention Shifts," *Proceedings of the National Academy of Sciences of the United States of America* 112, no. 27 (Jul. 2015): 8457–62.

44 M. Ertl, M. Hildebrandt, K. Ourina, G. Leicht, and C. Mulert, "Emotion Regulation by Cognitive Reappraisal—The Role of Frontal Theta Oscillations," *NeuroImage* 81 (Nov. 2013): 412–21.

45 R. F. Helfrich, T. R. Schneider, S. Rach, S. A. Trautmann-Lengsfeld, A. K. Engel, et al., "Entrainment of Brain Oscillations by Transcranial Alternating Current Stimulation," *Current Biology* 24 (2014): 333–39.

46 M. Bonnefond and O. Jensen, "Alpha Oscillations Serve to Protect Working Memory Maintenance against Anticipated Distracters," *Current Biology* 22 (2012): 1969–74.

47 C. F. Lavallee, S. A. Koren, and M. A. Persinger, "A Quantitative Electroencephalographic Study of Meditation and Binaural Beat Entrainment," *Journal of Alternative and Complementary Medicine* 17, no. 4 (Apr. 2011): 351–55.

48 L. Chaieb, E. C. Wilpert, T. P. Reber, and J. Fell, "Auditory Beat Stimulation and its Effects on Cognition and Mood States," *Frontiers in Psychiatry* 6 (May 2015): 70.

49 H. Wahbeh, C. Calabrese, H. Zwickey, and D. Zajdel, "Binaural Beat Technology in Humans: A Pilot Study to Assess Neuropsychologic, Physiologic, and Electroencephalographic Effects," *Journal of Alternative and Complementary Medicine* 13, no. 2 (Mar. 2007): 199–206.

50 R. Padmanabhan, A. J. Hildreth, and D. Laws, "A Prospective, Randomised, Controlled Study Examining Binaural Beat Audio and Preoperative Anxiety in Patients Undergoing General Anaesthesia for Day Case Surgery," *Anaesthesia* 60, no. 9 (Sept. 2005): 874–77.

51 P. A. McConnell, B. Froeliger, E. L. Garland, J. C. Ives, and G. A. Sforzo, "Auditory Driving of the Autonomic Nervous System: Listening to Theta-frequency Binaural Beats Post-exercise Increases Parasympathetic Activation and Sympathetic Withdrawal," *Frontiers in Psychology* 14, no. 5 (Nov. 2014): 1248.

52 K. Unno, K. Iguchi, N. Tanida, K. Fujitani, N. Takamori, H. Yamamoto, N. Ishii, H. Nagano, T. Nagashima, A. Hara, K. Shimoi, and M. Hoshino, "Ingestion of Theanine, an Amino Acid in Tea, Suppresses Psychosocial Stress in Mice," *Experimental Physiology* 98, no. 1 (Jan. 2013): 290–303.

53 A. L. Lardner, "Neurobiological Effects of the Green Tea Constituent Theanine and its Potential Role in the Treatment of Psychiatric and Neurodegenerative Disorders," *Nutritional Neuroscience* 17, no. 4 (Jul. 2014): 145–55.

54 D. A. Camfield, C. Stough, J. Farrimond, and A. B. Scholey, "Acute Effects of Tea Constituents L-theanine, Caffeine, and Epigallocate-

chin Gallate on Cognitive Function and Mood: A Systematic Review and Meta-analysis," *Nutrition Reviews* 72, no. 8 (Aug. 2014): 507–22.

55 S. Borgwardt, F. Hammann, K. Scheffler, M. Kreuter, J. Drewe, and C. Beglinger, "Neural Effects of Green Tea Extract on Dorsolateral Prefrontal Cortex," *European Journal of Clinical Nutrition* 66, no. 11 (Nov. 2012): 1187–92.

56 http://www.thenational.ae/news/peacefulness-through-a-bowl-of-tea.

CHAPTER 3

1 N. Skoluda, J. Strahler, W. Schlotz, L. Niederberger, S. Marques, S. Fischer, M. V. Thoma, C. Spoerri, U. Ehlert, and U. M. Nater, "Intraindividual Psychological and Physiological Responses to Acute Laboratory Stressors of Different Intensity," *Psychoneuroendocrinology* 51 (Jan. 2015): 227–36.

2 J. LeMoult and J. Joormann, "Depressive Rumination Alters Cortisol Decline in Major Depressive Disorder," *Biological Psychology* 100 (Jul. 2014): 50–55.

3 J. F. Brosschot, W. Gerin, and J. F. Thayer, "Worry and Health: The Perseverative Cognition Hypothesis," *Journal of Psychosomatic Research* 60 (2006): 113–24.

4 S. Nolen-Hoeksema, "The Role of Rumination in Depressive Disorders and Mixed Anxiety/Depressive Symptoms," *Journal of Abnormal Psychology* 109, no. 3 (2000): 504–11.

5 J. Morrow and S. Nolen-Hoeksema, "Effects of Responses to Depression on the Remediation of Depressive Affect," *Journal of Personality and Social Psychology* 58, no. 3 (1990): 519–27.

6 M. Glynn, N. Christenfeld, and W. Gerin, "The Role of Rumination in Recovery from Reactivity: Cardiovascular Consequences of Emotional States," *Psychosomatic Medicine* 64, no. 5 (2002): 714–26.

7 W. Gerin, K. W. Davidson, N. J. S. Christenfeld, T. Goyal, and J. E. Schwartz, "The Role of Angry Rumination and Distraction in Blood Pressure Recovery from Emotional Arousal," *Psychosomatic Medicine* 68 (2006): 64–72.

8 B. L. Key, T. S. Campbell, S. L. Bacon, and W. Gerin, "The Influence of Trait and State Rumination on Cardiovascular Recovery from a Negative Emotional Stressor," *Journal of Behavioral Medicine* 31 (2008): 237–48.

9 D. Gianferante, M. V. Thoma, L. Hanlin, et al., "Post-stress Rumination Predicts HPA Axis Responses to Repeated Acute Stress," *Psychoneuroendocrinology* 49 (2014): 244–52, doi: 10.1016/j.psyneuen.2014.07.021.

10 R. B. Price, B. Paul, W. Schneider, and G. J. Siegle, "Neural Correlates of Three Neurocognitive Intervention Strategies: A Preliminary Step Towards Personalized Treatment for Psychological Disorders," *Cognitive Therapy and Research* 37, no. 4 (Aug. 2013): 657–72.

11 J. Skorka-Brown, J. Andrade, and J. May, "Playing 'Tetris' Reduces the Strength, Frequency and Vividness of Naturally Occurring Cravings," *Appetite* 76 (May 2014): 161–65.

12 E. E. Hill, E. Zack, C. Battaglini, M. Viru, A. Viru, and A. C. Hackney, "Exercise and Circulating Cortisol Levels: The Intensity Threshold Effect," *Journal of Endocrinological Investigation* 31, no. 7 (Jul. 2008): 587–91.

13 E. Puterman, A. O'Donovan, N. E. Adler, A. J. Tomiyama, M. Kemeny, O. M. Wolkowitz, and E. Epel, "Physical Activity Moderates Effects of Stressor-induced Rumination on Cortisol Reactivity," *Psychosomatic Medicine* 73, no. 7 (Sept. 2011): 604–11.

14 B. Oneda, K. C. Ortega, J. L. Gusmão, T. G. Araújo, and D. Mion, Jr., "Sympathetic Nerve Activity is Decreased during Device-guided Slow Breathing," *Hypertension Research* 33, no. 7 (Jul. 2010): 708–12.

15 D. Harada, H. Asanoi, J. Takagawa, H. Ishise, H. Ueno, Y. Oda, Y. Goso, S. Joho, and H. Inoue, "Slow and Deep Respiration Suppresses Steady-state Sympathetic Nerve Activity in Patients with Chronic Heart Failure: From Modeling to Clinical Application," *American Journal of Physiology-Heart and Circulatory Physiology* 307, no. 8 (Oct. 2014): H1159–68.

16 R. Padmanabhan, A. J. Hildreth, and D. Laws, "A Prospective, Randomised, Controlled Study Examining Binaural Beat Audio and Pre-operative Anxiety in Patients Undergoing General Anaesthesia for Day Case Surgery," *Anaesthesia* 60 (2005): 874–77.

17 P. A. McConnell, B. Froeliger, E. L. Garland, J. C. Ives, and G. A. Sforzo, "Auditory Driving of the Autonomic Nervous System: Listening to Theta-frequency Binaural Beats Post-exercise Increases Parasympathetic Activation and Sympathetic Withdrawal," *Frontiers in Psychology* 14, no. 5 (Nov. 2014): 1248.

18 B. Gingras, G. Pohler, and W. T. Fitch, "Exploring Shamanic Journeying: Repetitive Drumming with Shamanic Instructions Induces

Specific Subjective Experiences but No Larger Cortisol Decrease Than Instrumental Meditation Music," *PLoS ONE* 9, no. 7 (Jul. 2014): e102103.

19 E. Largo-Wight, B. K. O'Hara, and W. W. Chen, "The Efficacy of a Brief Nature Sound Intervention on Muscle Tension, Pulse Rate, and Self-reported Stress: Nature Contact Micro-break in an Office or Waiting Room," *HERD* 10, no. 1 (Oct. 2016): 45–51.

20 M. Annerstedt, P. Jönsson, M. Wallergård, G. Johansson, B. Karlson, P. Grahn, A. M. Hansen, and P. Währborg, "Inducing Physiological Stress Recovery with Sounds of Nature in a Virtual Reality Forest—Results from a Pilot Study," *Physiology & Behavior* 118 (Jun. 2013): 240–50.

21 D. K. Brown, J. L. Barton, and V. F. Gladwell, "Viewing Nature Scenes Positively Affects Recovery of Autonomic Function Following Acute-Mental Stress," *Environmental Science & Technology* 47, no. 11 (2013): 5562–69.

22 V. F. Gladwell, D. K. Brown, J. L. Barton, M. P. Tarvainen, P. Kuoppa, J. Pretty, J. M. Suddaby, and G. R. Sandercock, "The Effects of Views of Nature on Autonomic Control," *European Journal of Applied Physiology* 112, no. 9 (Sept. 2012): 3379–86.

23 S. Dong and T. J. Jacob, "Combined Non-adaptive Light and Smell Stimuli Lowered Blood Pressure, Reduced Heart Rate and Reduced Negative Affect," *Physiology & Behavior* 156 (Mar. 2016): 94–105.

24 M. J. Henckens, G. A. van Wingen, M. Joels, and G. Fernandez, "Time-dependent Corticosteroid Modulation of Prefrontal Working Memory Processing," *Proceedings of the National Academy of Sciences of the United States of America* 108 (2011): 5801–06.

25 K. Imai, H. Sato, M. Hori, H. Kusuoka, H. Ozaki, H. Yokoyama, H. Takeda, M. Inoue, and T. Kamada, "Vagally Mediated Heart Rate Recovery after Exercise is Accelerated in Athletes but Blunted in Patients with Chronic Heart Failure," *Journal of the American College of Cardiology* 24, no. 6 (Nov. 1994): 1529–35.

26 T. Otsuki, S. Maeda, M. Iemitsu, Y. Saito, Y. Tanimura, J. Sugawara, R. Ajisaka, and T. Miyauchi, "Postexercise Heart Rate Recovery Accelerates in Strength-trained Athletes," *Medicine & Science in Sports & Exercise* 39, no. 2 (Feb. 2007): 365–70.

27 M. Nakamura, K. Hayashi, K. Aizawa, N. Mesaki, and I. Kono, "Effects of Regular Aerobic Exercise on Post-exercise Vagal Reactivation in Young Female," *European Journal of Sport Science* 13, no. 6 (2013): 674–80.

28 S. Seiler, O. Haugen, and E. Kuffel, "Autonomic Recovery after Exercise in Trained Athletes: Intensity and Duration Effects," *Medicine & Science in Sports & Exercise* 39, no. 8 (Aug. 2007): 1366–73.

29 J. Sugawara, H. Murakami, S. Maeda, S. Kuno, and M. Matsuda, "Change in Post-exercise Vagal Reactivation with Exercise Training and Detraining in Young Men," *European Journal of Applied Physiology* 85, nos. 3–4 (Aug. 2001): 259–63.

30 U. Rimmele, R. Seiler, B. Marti, P. H. Wirtz, U. Ehlert, and M. Heinrichs, "The Level of Physical Activity Affects Adrenal and Cardiovascular Reactivity to Psychosocial Stress," *Psychoneuroendocrinology* 34 (2009): 190–98.

31 E. Zschucke, B. Renneberg, F. Dimeo, T. Wüstenberg, and A. Ströhle, "The Stress-buffering Effect of Acute Exercise: Evidence for HPA Axis Negative Feedback," *Psychoneuroendocrinology* 51 (Jan. 2015): 414–25.

32 T. Baghurst and B. C. Kelley, "An examination of stress in college students over the course of a semester," *Health Promotion Practice* 15, no. 3 (May 2014): 438–47.

33 P. Kaikkonen, A. Nummela, and H. Rusko, "Heart Rate Variability Dynamics during Early Recovery after Different Endurance Exercises," *European Journal of Applied Physiology* 102 (2007): 79–86.

34 P. Kaikkonen, H. Rusko, and K. Martinma, "Post-exercise Heart Rate Variability of Endurance Athletes after Different High-intensity Exercises," *Scandinavian Journal of Medicine & Science in Sports* 18 (2008): 511–19.

35 M. M. Tanskanen, H. Kyröläinen, A. L. Uusitalo, J. Huovinen, J. Nissilä, H. Kinnunen, M. Atalay, and K. Häkkinen, "Serum Sex Hormone-binding Globulin and Cortisol Concentrations Are Associated with Overreaching during Strenuous Military Training," *Journal of Strength and Conditioning Research* 25, no. 3 (Mar. 2011): 787–97.

36 J. L. Abelson, T. M. Erickson, S. E. Mayer, J. Crocker, H. Briggs, N. L. Lopez-Duran, and I. Liberzon, "Brief Cognitive Intervention Can Modulate Neuroendocrine Stress Responses to the Trier Social Stress Test: Buffering Effects of a Compassionate Goal Orientation," *Psychoneuroendocrinology* 44 (Jun. 2014): 60–70.

37 A. A. Mohammadi, S. Jazayeri, K. Khosravi-Darani, Z. Solati, N. Mohammadpour, Z. Asemi, Z. Adab, M. Djalali, M. Tehrani-Doost, M. Hosseini, and S. Eghtesadi, "The Effects of Probiotics on Mental Health and Hypothalamic-pituitary-adrenal Axis: A

Randomized, Double-blind, Placebo-controlled Trial in Petrochemical Workers," *Nutritional Neuroscience* 19, no. 9 (Apr. 2015): 387–95.

38 T. Backes, P. Horvath, and K. Kazial, "Salivary Alpha Amylase and Salivary Cortisol Response to Fluid Consumption in Exercising Athletes," *Biology of Sport* 32, no. 4 (2015): 275–80.

39 B. R. Ely, K. J. Sollanek, S. N. Cheuvront, H. R. Lieberman, and R. W. Kenefick, "Hypohydration and Acute Thermal Stress Affect Mood State but Not Cognition or Dynamic Postural Balance," *European Journal of Applied Physiology* 113, no. 4 (Apr. 2013): 1027–34.

40 R. Micha, P. J. Rogers, and M. Nelson, "Glycaemic Index and Glycaemic Load of Breakfast Predict Cognitive Function and Mood in School Children: A Randomised Controlled Trial," *British Journal of Nutrition* 106, no. 10 (Nov. 2011): 1552–61.

41 A. R. Allen, L. R. Gullixson, S. C. Wolhart, S. L. Kost, D. R. Schroeder, and J. H. Eisenach, "Dietary Sodium Influences the Effect of Mental Stress on Heart Rate Variability: A Randomized Trial in Healthy Adults," *Journal of Hypertension* 32, no. 2 (Feb. 2014): 374–82.

42 J. B. Beckman, M. L. Stock, and T. Marcus, "Need to Belong, Not Rejection Sensitivity, Moderates Cortisol Response, Self-Reported Stress, and Negative Affect Following Social Exclusion," *Journal of Social Psychology* 156, no. 2 (2016): 131–38.

43 M. T. Bowen and I. S. McGregor, "Oxytocin and Vasopressin Modulate the Social Response to Threat: A Preclinical Study," *International Journal of Neuropsychopharmacology* 17, no. 10 (Oct. 2014): 1621–33.

44 M. Heinrichs, T. Baumgartner, C. Kirschbaum, and U. Ehlert, "Social Support and Oxytocin Interact to Suppress Cortisol and Subjective Responses to Psychosocial Stress," *Biological Psychiatry* 54 (2003): 1389–98.

45 C. Crockford, T. Deschner, T. E. Ziegler, and R. M. Wittig, "Endogenous Peripheral Oxytocin Measures Can Give Insight into the Dynamics of Social Relationships: A Review," *Frontiers in Behavioral Neuroscience* 8 (2014): 68.

46 K. M. Grewen, S. S. Girdler, J. Amico, and K. C. Light, "Effects of Partner Support on Resting Oxytocin, Cortisol, Norepinephrine, and Blood Pressure before and after Warm Partner Contact," *Psychosomatic Medicine* 67, no. 4 (Jul.–Aug. 2005): 531–38.

47 S. Ogawa, S. Kudo, Y. Kitsunai, and S. Fukuchi, "Increase in Oxytocin Secretion at Ejaculation in Male," *Clinical Endocrinology* 13 (1980): 95–97.

48 M. S. Carmichael, R. Humbert, J. Dixen, G. Palmisano, W. Green-leaf, and J. M. Davidson, "Plasma Oxytocin Increases in the Human Sexual Response," *Journal of Clinical Endocrinology & Metabolism* 64 (1987): 27–31.

49 R. A. Turner, M. Altemus, T. Enos, B. Cooper, and T. McGuinness, "Preliminary Research on Plasma Oxytocin in Normal Cycling Women: Investigating Emotion and Interpersonal Distress," *Psychiatry* 62 (1999): 97–113.

50 M. H. Rapaport, P. Schettler, and C. Bresee, "A Preliminary Study of the Effects of Repeated Massage on Hypothalamic-pituitary-adrenal and Immune Function in Healthy Individuals: A Study of Mechanisms of Action and Dosage," *Journal of Alternative and Complementary Medicine* 18, no. 8 (Aug. 2012): 789–97.

51 http://www.businessinsider.com/professional-cuddler-2014-7.

52 T. L. Kraft and S. D. Pressman, "Grin and Bear It: The Influence of Manipulated Facial Expression on the Stress Response," *Psychological Science* 23, no. 11 (2012): 1372–78.

53 S. S. Dickerson and M. E. Kemeny, "Acute Stressors and Cortisol Responses: A Theoretical Integration and Synthesis of Laboratory Research," *Psychological Bulletin* 130 (2004): 355–91.

54 Y. Yamanaka, S. Hashimoto, N. N. Takasu, Y. Tanahashi, S. Y. Nishide, S. Honma, and K. Honma, "Morning and Evening Physical Exercise Differentially Regulate the Autonomic Nervous System during Nocturnal Sleep in Humans," *American Journal of Physiology—Regulatory, Integrative and Comparative Physiology* 309, no. 9 (Nov. 2015): R1112–21.

55 S. Kuwahata, M. Miyata, S. Fujita, T. Kubozono, T. Shinsato, Y. Ikeda, S. Hamasaki, T. Kuwaki, and C. Tei, "Improvement of Autonomic Nervous Activity by Waon Therapy in Patients with Chronic Heart Failure," *Journal of Cardiology* 57, no. 1 (Jan. 2011): 100–6.

56 Y. Soejima, T. Munemoto, A. Masuda, Y. Uwatoko, M. Miyata, and C. Tei, "Effects of Waon Therapy on Chronic Fatigue Syndrome: A Pilot Study," *Internal Medicine* 54, no. 3 (2015): 333–38.

57 G. Kanji, M. Weatherall, R. Peter, G. Purdie, and R. Page, "Efficacy of Regular Sauna Bathing for Chronic Tension-type Headache: A Randomized Controlled Study," *Journal of Alternative and Complementary Medicine* 21, no. 2 (Feb. 2015): 103–9.

58 J. H. Fowler and N. A. Christakis, "Dynamic Spread of Happiness in a Large Social Network: Longitudinal Analysis over 20 Years in the Framingham Heart Study," *British Medical Journal* 337 (Dec. 2008): a2338.

59 O. A. Coubard. "An Integrative Model for the Neural Mechanism of Eye Movement Desensitization and Reprocessing (EMDR)," *Frontiers in Behavioral Neuroscience* 10 (2016): 52.

60 https://www.youtube.com/watch?v=nylajeG6uFY. Retrieved August 5, 2016.

61 C. Acarturk, E. Konuk, M. Cetinkaya, I. Senay, M. Sijbrandij, B. Gulen, and P. Cuijpers, "The Efficacy of Eye Movement Desensitization and Reprocessing for Post-traumatic Stress Disorder and Depression among Syrian Refugees: Results of a Randomized Controlled Trial," *Psychological Medicine* 46, no. 12 (Sept. 2016): 2583–93.

62 M. Maroufi, S. Zamani, Z. Izadikhah, M. Marofi, and P. O'Connor, "Investigating the Effect of Eye Movement Desensitization and Reprocessing (EMDR) on Postoperative Pain Intensity in Adolescents Undergoing Surgery: A Randomized Controlled Trial," *Journal of Advanced Nursing* 72, no. 9 (Sept. 2016): 2207–17.

63 R. N. McLay, J. A. Webb-Murphy, S. F. Fesperman, E. M. Delaney, S. K. Gerard, S. C. Roesch, B. J. Nebeker, I. Pandzic, E. A. Vishnyak, and S. L. Johnston, "Outcomes from Eye Movement Desensitization and Reprocessing in Active-Duty Service Members with Posttraumatic Stress Disorder," *Psychological Trauma* 8, no. 6 (Mar. 2016): 702–8.

CHAPTER 4

1 M. Banasr, G. W. Valentine, X. Y. Li, S. L. Gourley, J. R. Taylor, and R. S. Duman, "Chronic Unpredictable Stress Decreases Cell Proliferation in the Cerebral Cortex of the Adult Rat." *Biological Psychiatry* 62, no. 5 (Sept. 2007): 496–504.

2 E. Blix, A. Perski A, H. Berglund, and I. Savic, "Long-term Occupational Stress is Associated with Regional Reductions in Brain Tissue Volumes," *PLoS One* 8, no. 6 (Jun. 2013): e64065.

3 C. D. Gipson and M. F. Olive, "Structural and Functional Plasticity of Dendritic Spines—Root or Result of Behavior?" *Genes, Brain and Behavior 2017* 16, no. 1 (Jan. 2017): 101–17.

4 V. Pinto, J. C. Costa, P. Morgado, C. Mota, A. Miranda, F. V. Bravo, T. G. Oliveira, J. J. Cerqueira, and N. Sousa, "Differential Impact of Chronic Stress Along the Hippocampal Dorsal-ventral Axis," *Brain Structure and Function* 220, no. 2 (Mar. 2015): 1205–12.

5 A. Ashokan, A. Hegde, and R. Mitra, "Short-term Environmental Enrichment is Sufficient to Counter Stress-induced Anxiety and

Associated Structural and Molecular Plasticity in Basolateral Amygdala," *Psychoneuroendocrinology* 69 (Jul. 2016): 189–96.

6   A. N. Sharma, E. da Costa, B. F. Silva, J. C. Soares, A. F. Carvalho, and J. Quevedo, "Role of Trophic Factors GDNF, IGF-1 and VEGF in Major Depressive Disorder: A Comprehensive Review of Human Studies," *Journal of Affective Disorders* 197 (Jun. 2016): 9–20.

7   M. E. Maynard, C. Chung, A. Comer, K. Nelson, J. Tran, N. Werries, E. A. Barton, M. Spinetta, and J. L. Leasure, "Ambient Temperature Influences the Neural Benefits of Exercise," *Behavioural Brain Research* 299 (Feb. 2016): 27–31.

8   N. M. Vega-Rivera, L. Ortiz-López, A. Gómez-Sánchez, J. Oikawa-Sala, E. M. Estrada-Camarena, and G. B. Ramírez-Rodríguez, "The Neurogenic Effects of an Enriched Environment and its Protection against the Behavioral Consequences of Chronic Mild Stress Persistent after Enrichment Cessation in Six-month-old Female Balb/ C Mice," *Behavioural Brain Research* 301 (Mar. 2016): 72–83.

9   E. Castilla-Ortega, C. Rosell-Valle, C. Pedraza, F. Rodríguez de Fonseca, G. Estivill-Torrús, and L. J. Santín, "Voluntary Exercise Followed by Chronic Stress Strikingly Increases Mature Adult-born Hippocampal Neurons and Prevents Stress-induced Deficits in 'What-when-where' Memory," *Neurobiology of Learning and Memory* 109 (Mar. 2014): 62–73.

10   H. van Praag, G. Kempermann, and F. H. Gage, "Running Increases Cell Proliferation and Neurogenesis in the Adult Mouse Dentate Gyrus," *Nature Neuroscience* 2, no. 3 (Mar. 1999): 266–70.

11   S. A. Neeper, F. Gomez-Pinilla, J. Choi, and C. W. Cotman, "Physical Activity Increases mRNA for Brain-derived Neurotrophic Factor and Nerve Growth Factor in Rat Brain," *Brain Research* 726 (1996): 49–56.

12   M. S. Nokia, S. Lensu, J. P. Ahtiainen, P. P. Johansson, L. G. Koch, S. L. Britton, and H. Kainulainen, "Physical Exercise Increases Adult Hippocampal Neurogenesis in Male Rats Provided it is Aerobic and Sustained," *Journal of Physiology* 594, no. 7 (Apr. 2016): 1855–73.

13   K. Inoue, M. Okamoto, J. Shibato, M. C. Lee, T. Matsui, R. Rakwal, and H. Soya, "Long-term Mild, Rather than Intense, Exercise Enhances Adult Hippocampal Neurogenesis and Greatly Changes the Transcriptomic Profile of the Hippocampus," *PLoS ONE* 10, no. 6 (Jun. 2015): e0128720.

14   A. Dinoff, N. Herrmann, W. Swardfager, C. S. Liu, C. Sherman, S. Chan, and K. L. Lanctôt, "The Effect of Exercise Training on Resting

Concentrations of Peripheral Brain-Derived Neurotrophic Factor (BDNF): A Meta-Analysis," *PLoS ONE* 11, no. 9 (Sept. 2016): e0163037.

15 K. I. Erickson, M. W. Voss, R. S. Prakash, C. Basak, A. Szabo, L. Chaddock, J. S. Kim, S. Heo, H. Alves, S. M. White, T. R. Wojcicki, E. Mailey, V. J. Vieira, S. A. Martin, B. D. Pence, J. A. Woods, E. McAuley, and A. F. Kramer, "Exercise Training Increases Size of Hippocampus and Improves Memory," *Proceedings of the National Academy of Sciences of the United States of America* 108, no. 7 (Feb. 2011): 3017–22.

16 C. H. Hillman, K. I. Erickson, and A. F. Kramer, "Be Smart, Exercise Your Heart: Exercise Effects on Brain and Cognition," *Nature Reviews Neuroscience* 9, no. 1 (Jan. 2008): 58–65.

17 M. Zagaar, A. Dao, A. Levine, I. Alhaider, and K. Alkadhi, "Regular Exercise Prevents Sleep Deprivation Associated Impairment of Long-term Memory and Synaptic Plasticity in the CA1 Area of the Hippocampus," *Sleep* 36, no. 5 (May 2013): 751–61.

18 T. Kobilo, Q. R. Liu, K. Gandhi, et al., "Running is the Neurogenic and Neurotrophic Stimulus in Environmental Enrichment," *Learning & Memory* 18 (2011): 605–9.

19 G. Wagner, M. Herbsleb, F. Cruz, A. Schumann, F. Brünner, C. Schachtzabel, A. Gussew, C. Puta, S. Smesny, H. W. Gabriel, J. R. Reichenbach, and K. J. Bär, "Hippocampal Structure, Metabolism, and Inflammatory Response after a 6-week Intense Aerobic Exercise in Healthy Young Adults: A Controlled Trial," *Journal of Cerebral Blood Flow & Metabolism* 35, no.10 (Oct. 2015): 1570–78.

20 C. J. Heyser, E. Masliah, A. Samimi, I. L. Campbell, and L. H. Gold, "Progressive Decline in Avoidance Learning Paralleled by Inflammatory Neurodegeneration in Transgenic Mice Expressing Interleukin 6 in the Brain," *Proceedings of the National Academy of Sciences of the United States of America* 94 (1997): 1500–05.

21 D. L. Gruol, "IL-6 Regulation of Synaptic Function in the CNS," *Neuropharmacology* 96 (2015): 42–54, doi:10.1016/j.neuropharm.2014.10.023.

22 M. A. Ajmone-Cat, E. Cacci, and L. Minghetti, "Non Steroidal Antiinflammatory Drugs and Neurogenesis in the Adult Mammalian Brain," *Current Pharmaceutical Design* 14, no. 14 (2008): 1435–42.

23 K. Inoue, M. Okamoto, J. Shibato, M. C. Lee, T. Matsui, R. Rakwal, and H. Soya, "Long-term Mild, Rather than Intense, Exercise Enhances Adult Hippocampal Neurogenesis and Greatly Changes the Transcriptomic Profile of the Hippocampus," *PLoS ONE* 10, no. 6 (Jun. 2015): e0128720.

24 M. E. Maynard, C. Chung, A. Comer, K. Nelson, J. Tran, N. Werries, E. A. Barton, M. Spinetta, and J. L. Leasure, "Ambient Temperature Influences the Neural Benefits of Exercise," *Behavioural Brain Research* 299 (Feb. 2016): 27–31.

25 G. Umschweif, D. Shabashov, A. G. Alexandrovich, V. Trembovler, M. Horowitz, and E. Shohami, "Neuroprotection after Traumatic Brain Injury in Heat-acclimated Mice Involves Induced Neurogenesis and Activation of Angiotensin Receptor Type 2 Signaling," *Journal of Cerebral Blood Flow & Metabolism* 34, no. 8 (2014): 1381–90.

26 V. Bhagya, B. N. Srikumar, J. Veena, and B. S. Shankaranarayana Rao, "Short-term Exposure to Enriched Environment Rescues Chronic Stress-induced Impaired Hippocampal Synaptic Plasticity, Anxiety, and Memory Deficits," *Journal of Neuroscience Research* (Nov. 2016) (*epub ahead of print*) doi:10.1002/jnr.23992.

27 T. Novkovic, T. Mittmann, and D. Manahan-Vaughan, "BDNF Contributes to the Facilitation of Hippocampal Synaptic Plasticity and Learning Enabled by Environmental Enrichment," *Hippocampus* 25, no. 1 (Jan. 2015): 1–15.

28 J. Aarse, S. Herlitze, and D. Manahan-Vaughan, "The Requirement of BDNF for Hippocampal Synaptic Plasticity is Experience-dependent," *Hippocampus* 26, no. 6 (Jun. 2016): 739–51.

29 A. Ashokan, A. Hegde, and R. Mitra, "Short-term Environmental Enrichment is Sufficient to Counter Stress-induced Anxiety and Associated Structural and Molecular Plasticity in Basolateral Amygdala," *Psychoneuroendocrinology* 69 (Jul. 2016): 189–96.

30 G. D. Clemenson and C. E. L. Stark, "Virtual Environmental Enrichment through Video Games Improves Hippocampal-Associated Memory," *Journal of Neuroscience* 35, no. 49 (Dec. 2015): 16116–25.

31 S. Seinfeld, H. Figueroa, J. Ortiz-Gil, and M. V. Sanchez-Vives, "Effects of Music Learning and Piano Practice on Cognitive Function, Mood and Quality of Life in Older Adults," *Frontiers in Psychology* 4 (2013): 810.

32 J. Oltmanns, B. Godde, A. H. Winneke, G. Richter, C. Niemann, C. Voelcker-Rehage, K. Schömann, and U. M. Staudinger, "Don't Lose Your Brain at Work—The Role of Recurrent Novelty at Work in Cognitive and Brain Aging," *Frontiers in Psychology* 8 (2017): 117.

33 A. Arslan-Ergul, A. T. Ozdemir, and M. M. Adams, "Aging, Neurogenesis, and Caloric Restriction in Different Model Organisms," *Aging and Disease* 4, no. 4 (Jun. 2013): 221–32.

34   A. V. Witte, M. Fobker, R. Gellner, S. Knecht, and A. Floel, "Caloric Restriction Improves Memory in Elderly Humans," *Proceedings of the National Academy of Sciences of the United States of America* 106 (2009): 1255–60.

35   A. K. E. Hornsby, Y. T. Redhead, D. J. Rees, et al., "Short-term Calorie Restriction Enhances Adult Hippocampal Neurogenesis and Remote Fear Memory in a Ghsr-dependent Manner," *Psychoneuroendocrinology* 63 (2016): 198–207.

36   L. L. Hurley, L. Akinfiresoye, E. Nwulia, A. Kamiya, A. A. Kulkarni, and Y. Tizabi, "Antidepressant-like Effects of Curcumin in WKY Rat Model of Depression is Associated with an Increase in Hippocampal BDNF," *Behavioural Brain Research* 239 (Feb. 2013): 27–30.

37   D. Liu, Z. Wang, Z. Gao, K. Xie, Q. Zhang, H. Jiang, and Q. Pang, "Effects of Curcumin on Learning and Memory Deficits, BDNF, and ERK Protein Expression in Rats Exposed to Chronic Unpredictable Stress," *Behavioural Brain Research* 271 (Sept. 2014): 116–21.

38   A. L. Lopresti, M. Maes, G. L. Maker, S. D. Hood, and P. D. Drummond, "Curcumin for the Treatment of Major Depression: A Randomised, Double-blind, Placebo-controlled Study," *Journal of Affective Disorders* 167 (2014): 368–75.

39   A. L. Lopresti and P. D. Drummond, "Efficacy of Curcumin, and a Saffron/Curcumin Combination for the Treatment of Major Depression: A Randomised, Double-blind, Placebo-controlled Study," *Journal of Affective Disorders* 207 (Jan. 2017): 188–96.

40   B. B. Aggarwal, W. Yuan, S. Li, and S. C. Gupta, "Curcumin-free Turmeric Exhibits Anti-inflammatory and Anticancer Activities: Identification of Novel Components of Turmeric," *Molecular Nutrition & Food Research* 57, no. 9 (Sept. 2013): 1529–42.

41   R. W. Johnson, "Feeding the Beast: Can Microglia in the Senescent Brain Be Regulated by Diet?" *Brain, Behavior, and Immunity* 43 (Jan. 2015): 1–8.

42   V. Chesnokova, R. N. Pechnick, and K. Wawrowsky, "Chronic Peripheral Inflammation, Hippocampal Neurogenesis, and Behavior," *Brain, Behavior, and Immunity* 58 (Nov. 2016): 1–8.

43   D. Csabai, K. Cseko, L. Szaiff, Z. Varga, A. Miseta, Z. Helyes, and B. Czéh, "Low Intensity, Long Term Exposure to Tobacco Smoke Inhibits Hippocampal Neurogenesis in Adult Mice," *Behavioural Brain Research* 302 (Apr. 2016): 44–52.

44   Y. H. Shih, S. F. Tsai, S. H. Huang, Y. T. Chiang, M. W. Hughes, S. Y. Wu, C. W. Lee, T. T. Yang, and Y. M. Kuo, "Hypertension Impairs

Hippocampus-related Adult Neurogenesis, CA1 Neuron Dendritic Arborization and Long-term Memory," *Neuroscience* 322 (Feb. 2016): 346–57.

CHAPTER 5

1   M. Bellesi, L. deVivo, G. Tononi, and C. Cirelli, "Effects of Sleep and Wake on Astrocytes: 896 Clues from Molecular and Ultrastructural Studies," *BMC Biology* 13, no. 66 (2015).

2   G. Bernardi, L. Cecchetti, F. Siclari, A. Buchmann, X. Yu, G. Handjaras, M. Bellesi, E. Ricciardi, S. R. Kecskemeti, B. A. Riedner, A. L. Alexander, R. M. Benca, M. F. Ghilardi, P. Pietrini, C. Cirelli, and G. Tononi, "Sleep Reverts Changes in Human Gray and White Matter Caused by Wake-dependent Training," *NeuroImage* 129 (Apr. 2016): 367–77.

3   A. A. Borbely, "The S-deficiency Hypothesis of Depression and the Two-Process Model of Sleep Regulation," *Pharmacopsychiatry* 20 (1987): 23–29.

4   A. Germain and M. Dretsch, "Sleep and Resilience—A Call for Prevention and Intervention," *Sleep* 39, no. 5 (2016): 963–65.

5   M. M. Menz, J. S. Rihm, and C. Büchel, "REM Sleep Is Causal to Successful Consolidation of Dangerous and Safety Stimuli and Reduces Return of Fear after Extinction," *Journal of Neuroscience* 36, no. 7 (Feb. 2016): 2148–60.

6   M. Razzoli, C. Karsten, J. M. Yoder, A. Bartolomucci, and W. C. Engeland, "Chronic Subordination Stress Phase Advances Adrenal and Anterior Pituitary Clock Gene Rhythms," *American Journal of Physiology—Regulatory, Integrative and Comparative Physiology* 307, no. 2 (Jul. 2014): R198–205.

7   Y. Tahara, T. Shiraishi, Y. Kikuchi, A. Haraguchi, D. Kuriki, H. Sasaki, H. Motohashi, T. Sakai, and S. Shibata, "Entrainment of the Mouse Circadian Clock by Sub-acute Physical and Psychological Stress," *Scientific Reports* 5 (Jun. 2015): 11417.

8   P. D. Mavroudis, S. A. Corbett, S. E. Calvano, and I. P. Androulakis, "Mathematical Modeling of Light-mediated HPA Axis Activity and Downstream Implications on the Entrainment of Peripheral Clock Genes," *Physiological Genomics* 46, no. 20 (Oct. 2014): 766–78.

9   P. Pezük, J. A. Mohawk, L. A. Wang, and M. Menaker, "Glucocorticoids as Entraining Signals for Peripheral Circadian Oscillators," *Endocrinology* 153, no. 10 (Oct. 2012): 4775–83.

10 P. D. Mavroudis, J. D. Scheff, S. E. Calvano, S. F. Lowry, and I. P. Androulakis, "Entrainment of Peripheral Clock Genes by Cortisol," *Physiological Genomics* 44, no. 11 (2012): 607–21.

11 L. D. Grandin, L. B. Alloy, and L. Y. Abramson, "The Social Zeitgeber Theory, Circadian Rhythms, and Mood Disorders: Review and Evaluation," *Clinical Psychology Review* 26, no. 6 (Oct. 2006): 679–94.

12 B. Etain, V. Milhiet, F. Bellivier, and M. Leboyer, "Genetics of Circadian Rhythms and Mood Spectrum Disorders," *European Neuropsychopharmacology: The Journal of the European College of Neuropsychopharmacology* 21, no. 4 (2011): S676–82.

13 P. Pevet and E. Challet, "Melatonin: Both Master Clock Output and Internal Time-giver in the Circadian Clocks Network," *Journal of Physiology–Paris* 105, nos. 4–6 (Dec. 2011): 170–82.

14 E. Grossman, M. Laudon, and N. Zisapel, "Effect of Melatonin on Nocturnal Blood Pressure: Meta-analysis of Randomized Controlled Trials," *Journal of Vascular Health and Risk Management* 7 (2011): 577–84.

15 F. Simko, T. Baka, L. Paulis, and R. J. Reiter, "Elevated Heart Rate and Nondipping Heart Rate as Potential Targets for Melatonin: A Review," *Journal of Pineal Research* 61, no. 2 (Sept. 2016): 127–37.

16 M. D. Muller, C. L. Sauder, and C. A. Ray, "Melatonin Attenuates the Skin Sympathetic Nerve Response to Mental Stress," *American Journal of Physiology—Heart and Circulatory Physiology* 305, no. 9 (Nov. 2013): H1382–86.

17 N. Ruksee, W. Tongjaroenbuangam, T. Mahanam, and P. Govitrapong, "Melatonin Pretreatment Prevented the Effect of Dexamethasone Negative Alterations on Behavior and Hippocampal Neurogenesis in the Mouse Brain," *Journal of Steroid Biochemistry and Molecular Biology* 143 (Sept. 2014): 72–80.

18 F. Boulle, R. Massart, E. Stragier, et al., "Hippocampal and Behavioral Dysfunctions in a Mouse Model of Environmental Stress: Normalization by Agomelatine," *Translational Psychiatry* 4, no. 11 (2014): e485.

19 D. F. Kripke, J. A. Elliott, D. K. Welsh, and S. D. Youngstedt, "Photoperiodic and Circadian Bifurcation Theories of Depression and Mania," *F1000Research* 4 (May 2015): 107.

20 T. E. Henriksen, S. Skrede, O. B. Fasmer, H. Schoeyen, I. Leskauskaite, J. Bjørke-Bertheussen, J. Assmus, B. Hamre, J. Grønli, and A. Lund, "Blue-blocking Glasses as Additive Treatment for Mania:

A Randomized Placebo-controlled Trial." *Bipolar Disorders* 18, no. 3 (May 2016): 221–32.

21   A. J. Lewy, "The Dim Light Melatonin Onset, Melatonin Assays and Biological Rhythm Research in Humans," *Biological Signals and Receptors* 8, nos. 1–2 (Jan. 1999): 79–83.

22   A. Wahnschaffe, S. Haedel, A. Rodenbeck, et al., "Out of the Lab and into the Bathroom: Evening Short-term Exposure to Conventional Light Suppresses Melatonin and Increases Alertness Perception," *International Journal of Molecular Sciences* 14, no. 2 (2013): 2573–89.

23   T. Kozaki, A. Kubokawa, R. Taketomi, and K. Hatae, "Light-induced Melatonin Suppression at Night after Exposure to Different Wavelength Composition of Morning Light," *Neuroscience Letters* 616 (Jan. 2016): 1–4.

24   V. Gabel, M. Maire, C. F. Reichert, S. L. Chellappa, C. Schmidt, V. Hommes, A. U. Viola, and C. Cajochen, "Effects of Artificial Dawn and Morning Blue Light on Daytime Cognitive Performance, Well-being, Cortisol and Melatonin Levels," *Chronobiology International* 30, no. 8 (Oct. 2013): 988–97.

25   C. A. Czeisler, R. E. Kronauer, J. S. Allan, J. F. Duffy, M. E. Jewett, E. N. Brown, and J. M. Ronda, "Bright Light Induction of Strong (Type 0) Resetting of the Human Circadian Pacemaker," *Science* 244, no. 4910 (Jun. 1989): 1328–33.

26   S. J. Kim, S. Benloucif, K. J. Reid, S. Weintraub, N. Kennedy, L. F. Wolfe, and P. C. Zee, "Phase-shifting Response to Light in Older Adults." *The Journal of Physiology* 592, Pt. 1 (2014): 189–202.

27   K. Dewan, S. Benloucif, K. Reid, L. F. Wolfe, and P. C. Zee, "Light-induced Changes of the Circadian Clock of Humans: Increasing Duration is More Effective than Increasing Light Intensity," *Sleep* 34, no. 5 (May 2011): 593–99.

28   M. Rüger, M. A. St. Hilaire, G. C. Brainard, S. B. Khalsa, R. E. Kronauer, C. A. Czeisler, et al., "Human Phase Response Curve to a Single 6.5 h Pulse of Short-wavelength Light," *Journal of Physiology* 591 (2013): 353–63.

29   K. Wada, S. Yata, O. Akimitsu, M. Krejci, T. Noji, M. Nakade, H. Takeuchi, and T. Harada, "A Tryptophan-rich Breakfast and Exposure to Light with Low Color Temperature at Night Improve Sleep and Salivary Melatonin Level in Japanese Students," *Journal of Circadian Rhythms* 11 (May 2013): 4.

30   H. Fukushige, Y. Fukuda, M. Tanaka, K. Inami, K. Wada, Y. Tsumura, M. Kondo, T. Harada, T. Wakamura, and T. Morita, "Effects

of Tryptophan-rich Breakfast and Light Exposure during the Daytime on Melatonin Secretion at Night," *Journal of Physiological Anthropology* 33 (Nov. 2014): 33.

31 Y. Aguilera, M. Rebollo-Hernanz, T. Herrera, L. T. Cayuelas, P. Rodríguez-Rodríguez, Á. L. de Pablo, S. M. Arribas, and M. A. Martin-Cabrejas, "Intake of Bean Sprouts Influences Melatonin and Antioxidant Capacity Biomarker Levels in Rats," *Food & Function* 7, no. 3 (Mar. 2016): 1438–45.

32 G. Howatson, P. G. Bell, J. Tallent, B. Middleton, M. P. McHugh, and J. Ellis, "Effect of Tart Cherry Juice (Prunus cerasus) on Melatonin Levels and Enhanced Sleep Quality," *European Journal of Nutrition* (2012): 909–16.

33 A. Asher, A. Shabtay, A. Brosh, H. Eitam, R. Agmon, M. Cohen-Zinder, A. E. Zubidat, and A. Haim, "'Chrono-functional Milk': The Difference between Melatonin Concentrations in Night-milk versus Day-milk under Different Night Illumination Conditions," *Chronobiology International* 32, no. 10 (2015): 1409–16.

34 R. H. Goo, J. G. Moore, E. Greenberg, et al., "Circadian Variation in Gastric Emptying of Meals in Humans," *Gastroenterology* 93 (1987): 515–18.

35 S. S. Rao, P. Sadeghi, J. Beaty, et al., "Ambulatory 24-h Colonic Manometry in Healthy Humans," *American Journal of Physiology-Gastrointestinal and Liver Physiology* 280 (2001): G629–G639.

36 S. S. Han, R. Zhang, R. Jain, et al., "Circadian Control of Bile Acid Synthesis by a Klf15-fgf15 Axis," *Nature Communications* 6 (2015): 7231.

37 A. Zarrinpar, A. Chaix, S. Yooseph, et al., "Diet and Feeding Pattern Affect the Diurnal Dynamics of the Gut Microbiome," *Cell Metabolism* 20 (2014): 1006–17.

38 J. Lund, J. Arendt, S. M. Hampton, J. English, and L. M. Morgan, "Postprandial Hormone and Metabolic Responses amongst Shift Workers in Antarctica," *Journal of Endocrinology* 171, no. 3 (Dec. 2001): 557–64.

39 P. Rubio-Sastre, F. A. Scheer, P. Gómez-Abellán, J. A. Madrid, and M. Garaulet, "Acute melatonin administration in humans impairs glucose tolerance in both the morning and evening," *Sleep* 37, no. 10 (Oct. 2014): 1715-19.

40 C. Vollmers, S. Gill, L. DiTacchio, S. R. Pulivarthy, H. D. Le, and S. Panda, "Time of Feeding and the Intrinsic Circadian Clock Drive Rhythms in Hepatic Gene Expression," *Proceedings of the National*

Academy of Sciences of the United States of America 106, no. 50 (Dec. 2009): 21453–58.

41 A. Hirao, H. Nagahama, T. Tsuboi, M. Hirao, Y. Tahara, and S. Shibata, "Combination of Starvation Interval and Food Volume Determines the Phase of Liver Circadian Rhythm in Per2::Luc Knock-in Mice under Two Meals per Day Feeding," *American Journal of Physiology-Gastrointestinal and Liver Physiology* 299, no. 5 (2010): G1045–53.

42 H. Kuroda, Y. Tahara, K. Saito, N. Ohnishi, Y. Kubo, Y. Seo, et al., "Meal Frequency Patterns Determine the Phase of Mouse Peripheral Circadian Clocks," *Scientific Reports* 2 (2012): 711.

43 R. Hara, K. Wan, H. Wakamatsu, R. Aida, T. Moriya, M. Akiyama, et al., "Restricted Feeding Entrains Liver Clock without Participation of the Suprachiasmatic Nucleus," *Genes to Cells* 6, no. 3 (2001): 269–78.

44 T. Yoshizaki, Y. Tada, A. Hida, A. Sunami, Y. Yokoyama, J. Yasuda, A. Nakai, F. Togo, and Y. Kawano, "Effects of Feeding Schedule Changes on the Circadian Phase of the Cardiac Autonomic Nervous System and Serum Lipid Levels," *European Journal of Applied Physiology* 113, no. 10 (Oct. 2013): 2603–11.

45 C. Bandín, F. A. Scheer, A. J. Luque, V. Avila-Gandía, S. Zamora, J. A. Madrid, P. Gómez-Abellán, and M. Garaulet, "Meal Timing Affects Glucose Tolerance, Substrate Oxidation and Circadian-related Variables: A Randomized, Crossover Trial," *International Journal of Obesity* 39, no. 5 (May 2015): 828–33.

46 M. Hatori, C. Vollmers, A. Zarrinpar, L. DiTacchio, E. A. Bushong, S. Gill, M. Leblanc, A. Chaix, M. Joens, J. A. Fitzpatrick, M. H. Ellisman, and S. Panda, "Time-restricted Feeding without Reducing Caloric Intake Prevents Metabolic Diseases in Mice Fed a High-fat Diet," *Cell Metabolism* 15, no. 6 (Jun. 2012): 848–60.

47 D. Jakubowicz, M. Barnea, J. Wainstein, et al., "High Caloric Intake at Breakfast vs. Dinner Differentially Influences Weight Loss of Overweight and Obese Women," *Obesity* 21 (2013): 2504–12.

48 Y. Yamanaka, S. Hashimoto, N. N. Takasu, Y. Tanahashi, S. Y. Nishide, S. Honma, and K. Honma, "Morning and Evening Physical Exercise Differentially Regulate the Autonomic Nervous System during Nocturnal Sleep in Humans," *American Journal of Physiology— Regulatory, Integrative and Comparative Physiology* 309, no. 9 (Nov. 2015): R1112–21.

49   M. R. Ebben and A. J. Spielman, "The Effects of Distal Limb Warm-ing on Sleep Latency," *International Journal of Behavioral Medicine* 13, no. 3 (2006): 221–28.

50   T. M. Burke, R. R. Markwald, A. W. McHill, E. D. Chinoy, J. A. Snider, S. C. Bessman, C. M. Jung, J. S. O'Neill, and K. P. Wright Jr., "Effects of Caffeine on the Human Circadian Clock in vivo and in vitro," *Science Translational Medicine* 7, no. 305 (Sept. 2015): 305ra146.

51   http://www.mayoclinic.org/healthy-lifestyle/nutrition-and -healthy-eating/in-depth/caffeine/art-20045678.

52   H. Slama, G. Deliens, R. Schmitz, P. Peigneux, and R. Leproult, "After-noon Nap and Bright Light Exposure Improve Cognitive Flexibility Post Lunch," *PLoS ONE* 10, no. 5 (May 2015): e0125359.

53   H. Baek and B. K. Min, "Blue Light Aids in Coping with the Post-lunch Dip: An EEG Study," *Ergonomics* 58, no. 5 (2015): 803–10.

54   M. Takahashi and H. Arito, "Maintenance of Alertness and Perfor-mance by a Brief Nap after Lunch under Prior Sleep Deficit," *Sleep* 23, no. 6 (Sept. 2000): 813–19.

55   M. Takahashi, H. Fukuda, and H. Arito, "Brief Naps during Post-lunch Rest: Effects on Alertness, Performance, and Autonomic Bal-ance," *European Journal of Applied Physiology and Occupational Physiology* 78, no. 2 (Jul. 1998): 93–98.

56   J. A. Groeger, J. C. Lo, C. G. Burns, and D. J. Dijk, "Effects of Sleep Inertia after Daytime Naps Vary with Executive Load and Time of Day," *Behavioral Neuroscience* 125, no. 2 (Apr. 2011): 252–60.

57   M. Tamaki, A. Shirota, M. Hayashi, and T. Hori, "Restorative Effects of a Short Afternoon Nap (<30 min) in the Elderly on Subjective Mood, Performance and EEG Activity," *Sleep Research Online* 3 (2000): 131–39.

58   K. Müller, L. Libuda, A. M. Terschlüsen, and M. Kersting, "A Re-view of the Effects of Lunch on Adults' Short-term Cognitive Func-tioning," *Canadian Journal of Dietetic Practice and Research* 74, no. 4 (2013): 181–88.

59   L. A. Reyner, S. J. Wells, V. Mortlock, and J. A. Horne, "'Post-lunch' Sleepiness during Prolonged, Monotonous Driving—Effects of Meal Size," *Physiology & Behavior* 105, no. 4 (Feb. 2012): 1088–91.

CHAPTER 6

1   L. V. Borovikova, S. Ivanova, M. Zhang, H. Yang, G. I. Botchkina, L. R. Watkins, H. Wang, N. Abumrad, J. W. Eaton, and K. J. Tracey,

"Vagus Nerve Stimulation Attenuates the Systemic Inflammatory Response to Endotoxin," *Nature* 405 (2000): 458–62.

2   Ibid.

3   J. A. Sturgeon, A. Arewasikporn, M. A. Okun, M. C. Davis, A. D. Ong, and A. J. Zautra, "The Psychosocial Context of Financial Stress: Implications for Inflammation and Psychological Health," *Psychosomatic Medicine* 78, no. 2 (Nov. 2015): 134–43.

4   G. E. Miller, E. Chen, J. Sze, T. Marin, J. M. Arevalo, R. Doll, et al., "A Functional Genomic Fingerprint of Chronic Stress in Humans: Blunted Glucocorticoid and Increased NF-kappaB Signaling," *Biological Psychiatry* 64 (2008): 266–72.

5   Z. Visnovcova, D. Mokra, P. Mikolka, M. Mestanik, A. Jurko, M. Javorka, A. Calkovska, and I. Tonhajzerova, "Alterations in Vagal-immune Pathway in Long-lasting Mental Stress," *Advances in Experimental Medicine and Biology* 832 (2015): 45–50.

6   H. Besedovsky, A. del Rey, E. Sorkin, and C. A. Dinarello, "Immunoregulatory Feedback between Interleukin-1 and Glucocorticoid Hormones," *Science* 233, no. 4764 (Aug. 1986): 652–54.

7   When your immune system detects invading viruses and bacteria, it releases inflammatory agents. One such agent is interleukin-1 or IL-1. IL-1 triggers the release of the stress hormone cortisol by setting off a chain of commands along the stress axis.

8   R. Dantzer, J. C. O'Connor, G. G. Freund, R. W. Johnson, and K. W. Kelley, "From Inflammation to Sickness and Depression: When the Immune System Subjugates the Brain," *Nature Reviews Neuroscience* 9, no. 1 (Jan. 2008): 46–56.

9   P. W. Gold, "The Organization of the Stress System and its Dysregulation in Depressive Illness," *Molecular Psychiatry* 20, no. 1 (Feb. 2015): 32–47.

10 T. W. Pace, F. Hu, and A. H. Miller, "Cytokine-effects on Glucocorticoid Receptor Function: Relevance to Glucocorticoid Resistance and the Pathophysiology and Treatment of Major Depression," *Brain, Behavior, and Immunity* 21 (2007): 9–19.

11 C. L. Raison and A. H. Miller, "When Not Enough is Too Much: The Role of Insufficient Glucocorticoid Signaling in the Pathophysiology of Stress-related Disorders," *American Journal of Psychiatry* 160 (2003): 1554–65.

12 A. Bierhaus, J. Wolf, M. Andrassy, et al., "A Mechanism Converting Psychosocial Stress into Mononuclear Cell Activation," *Proceedings*

of the National Academy of Sciences of the United States of America 100, no. 4 (2003): 1920–25.

13 M. Iwata, K. T. Ota, X. Y. Li, F. Sakaue, N. Li, S. Dutheil, M. Banasr, V. Duric, T. Yamanashi, K. Kaneko, K. Rasmussen, A. Glasebrook, A. Koester, D. Song, K. A. Jones, S. Zorn, G. Smagin, and R. S. Duman, "Psychological Stress Activates the Inflammasome via Release of Adenosine Triphosphate and Stimulation of the Purinergic Type 2X7 Receptor," *Biological Psychiatry* 80, no. 1 (Dec. 2015): 12–22.

14 Y. Zhang, L. Liu, Y. Z. Liu, X. L. Shen, T. Y. Wu, T. Zhang, W. Wang, Y. X. Wang, and C. L. Jiang, "NLRP3 Inflammasome Mediates Chronic Mild Stress-Induced Depression in Mice via Neuroinflammation," *International Journal of Neuropsychopharmacology* 18, no. 8 (Jan. 2015), pii: pyv006.

15 M. Tartter, C. Hammen, J. E. Bower, P. A. Brennan, and S. Cole, "Effects of Chronic Interpersonal Stress Exposure on Depressive Symptoms Are Moderated by Genetic Variation at IL6 and IL1β in Youth," *Brain, Behavior, and Immunity* 46 (May 2015): 104–11.

16 M. Ozawa, M. Shipley, M. Kivimaki, A. Singh-Manoux, and E. J. Brunner, "Dietary Pattern, Inflammation and Cognitive Decline: The Whitehall II Prospective Cohort Study," *Clinical Nutrition* (Jan. 2016): pii: S0261-5614(16)00035-2.

17 C. L. Raison, L. Capuron, and A. H. Miller, "Cytokines Sing the Blues: Inflammation and the Pathogenesis of Depression," *Trends in Immunology* 27 (2006): 24–31.

18 C. L. Raison, R. E. Rutherford, B. J. Woolwine, et al., "A Randomized Controlled Trial of the Tumor Necrosis Factor Antagonist Infliximab for Treatment-resistant Depression: The Role of Baseline Inflammatory Biomarkers," *JAMA Psychiatry* 70 (2013): 31–41.

19 P. Yang, Z. Gao, H. Zhang, Z. Fang, C. Wu, H. Xu, and Q. J. Huang, "Changes in Proinflammatory Cytokines and White Matter in Chronically Stressed Rats," *Neuropsychiatric Disease and Treatment* 11 (Mar. 2015): 597–607.

20 B. Leonard and M. Maes, "Mechanistic Explanations How Cell-mediated Immune Activation, Inflammation, and Oxidative and Nitrosative Stress Pathways and Their Sequels and Concomitants Play a Role in the Pathophysiology of Unipolar Depression," *Neuroscience & Biobehavioral Reviews* 36 (2012): 764–85.

21 I. Goshen, T. Kreisel, O. Ben-Menachem-Zidon, T. Licht, J. Weidenfeld, T. Ben-Hur, et al., "Brain Interleukin-1 Mediates Chronic

Stress-induced Depression in Mice via Adrenocortical Activation and Hippocampal Neurogenesis Suppression," *Molecular Psychiatry* 13 (2008): 717–28.

22 J. L. Jankowsky and P. H. Patterson, "Cytokine and Growth Factor Involvement in Long-term Potentiation," *Molecular and Cellular Neuroscience* 14, nos. 4–5 (Oct.–Nov. 1999): 273–86.

23 G. Ravaglia, P. Forti, F. Maioli, N. Brunetti, M. Martelli, L. Servadei, L. Bastagli, M. Bianchin, and E. Mariani, "Serum C-reactive Protein and Cognitive Function in Healthy Elderly Italian Community Dwellers," *Journals of Gerontology Series A: Biological Sciences and Medical Sciences* 60, no. 8 (Aug. 2005): 1017–21.

24 J. N. Trollor, E. Smith, E. Agars, S. A. Kuan, B. T. Baune, L. Campbell, et al., "The Association between Systemic Inflammation and Cognitive Performance in the Elderly: The Sydney Memory and Ageing Study," *Age* 34 (2012): 1295–1308, 10.1007/s11357-011-9301-x.

25 S. M. Heringa, E. van den Berg, Y. D. Reijmer, G. Nijpels, C. D. Stehouwer, C. G. Schalkwijk, T. Teerlink, P. G. Scheffer, K. van den Hurk, L. J. Kappelle, J. M. Dekker, and G. J. Biessels, "Markers of Low-grade Inflammation and Endothelial Dysfunction Are Related to Reduced Information Processing Speed and Executive Functioning in an Older Population—the Hoorn Study," *Psychoneuroendocrinology* 40 (Feb. 2014): 108–18.

26 A. J. Ocon, "Caught in the Thickness of Brain Fog: Exploring the Cognitive Symptoms of Chronic Fatigue Syndrome," *Frontiers in Physiology* 4 (Apr. 2013): 63.

27 G. Lange, J. Steffener, D. B. Cook, B. M. Bly, C. Christodoulou, W. C. Liu, J. Deluca, and B. H. Natelson, "Objective Evidence of Cognitive Complaints in Chronic Fatigue Syndrome: A BOLD fMRI Study of Verbal Working Memory," *Neuroimage* 26, no. 2 (Jun. 2005): 513–24.

28 D. B. Cook, P. J. O'Connor, G. Lange, and J. Steffener, "Functional Neuroimaging Correlates of Mental Fatigue Induced by Cognition among Chronic Fatigue Syndrome Patients and Controls," *Neuroimage* 36, no. 1 (May 2007): 108–22.

29 G. Morris, M. Berk, K. Walder, and M. Maes, "Central Pathways Causing Fatigue in Neuro-inflammatory and Autoimmune Illnesses," *BMC Medicine* 13 (2015): 28, doi:10.1186/s12916-014-0259-2.

30 T. C. Theoharides, J. M. Stewart, S. Panagiotidou, and I. Melamed, "Mast Cells, Brain Inflammation and Autism," *European Journal of Pharmacology* (May 2015): pii: S0014-2999(15)00398-2.

31   T. W. Costantini, V. Bansal, C. Y. Peterson, et al., "Efferent Vagal Nerve Stimulation Attenuates Gut Barrier Injury after Burn: Modulation of Intestinal Occludin Expression," *Journal of Trauma* 68, no. 6 (2010): 1349–56.

32   J. S. Grigoleit, J. S. Kullmann, O. T. Wolf, F. Hammes, A. Wegner, S. Jablonowski, H. Engler, E. Gizewski, R. Oberbeck, and M. Schedlowski, "Dose-dependent Effects of Endotoxin on Neurobehavioral Functions in Humans," *PLoS ONE* 6 (2011): e28330.

33   J. S. Kullmann, J. S. Grigoleit, P. Lichte, P. Kobbe, C. Rosenberger, C. Banner, O. T. Wolf, H. Engler, R. Oberbeck, S. Elsenbruch, U. Bingel, M. Forsting, E. R. Gizewski, and M. Schedlowski, "Neural Response to Emotional Stimuli during Experimental Human Endotoxemia," *Human Brain Mapping* 34, no. 9 (Sept. 2013): 2217–27.

34   L. Giloteaux, J. K. Goodrich, W. A. Walters, S. M. Levine, R. E. Ley, and M. R. Hanson, "Reduced Diversity and Altered Composition of the Gut Microbiome in Individuals with Myalgic Encephalomyelitis/Chronic Fatigue Syndrome," *Microbiome* 4 (2016): 30.

35   T. Vanuytsel, S. van Wanrooy, H. Vanheel, C. Vanormelingen, S. Verschueren, E. Houben, S. Salim Rasoel, J. Tóth, L. Holvoet, R. Farré, L. van Oudenhove, G. Boeckxstaens, K. Verbeke, and J. Tack, "Psychological Stress and Corticotropin-releasing Hormone Increase Intestinal Permeability in Humans by a Mast Cell-dependent Mechanism," *Gut* 63, no. 8 (Aug. 2014): 1293–99.

36   M. Krzyzaniak, C. Peterson, W. Loomis, et al., "Postinjury Vagal Nerve Stimulation Protects against Intestinal Epithelial Barrier Breakdown," *Journal of Trauma* 70, no. 5 (2011): 1168–76, doi:10.1097/TA.0b013e318216f754.

37   J. M. Van Houten, R. J. Wessells, H. L. Lujan, and S. E. DiCarlo, "My Gut Feeling Says Rest: Increased Intestinal Permeability Contributes to Chronic Diseases in High-intensity Exercisers," *Medical Hypotheses* 85, no. 6 (Dec. 2015): 882–86.

38   X. Li, E. M. Kan, J. Lu, Y. Cao, R. K. Wong, A. Keshavarzian, and C. H. Wilder-Smith, "Combat-training Increases Intestinal Permeability, Immune Activation and Gastrointestinal Symptoms in Soldiers," *Alimentary Pharmacology and Therapeutics* 37, no. 8 (Apr. 2013): 799–809.

39   P. Schnohr, J. H. O'Keefe, J. L. Marott, P. Lange, and G. B. Jensen, "Dose of Jogging and Long-term Mortality," *Journal of the American College of Cardiology* 65 (2015): 411–19.

40 S. Alfonso-Loeches, J. Ureña-Peralta, M. J. Morillo-Bargues, U. Gómez-Pinedo, and C. Guerri, "Ethanol-induced TLR4/NLRP3 Neuroinflammatory Response in Microglial Cells Promotes Leukocyte Infiltration across the BBB," *Neurochemical Research* 41, nos. 1–2 (Feb. 2016): 193–209.

41 V. Purohit, J. C. Bode, C. Bode, D. A. Brenner, M. A. Choudhry, F. Hamilton, Y. J. Kang, A. Keshavarzian, R. Rao, R. B. Sartor, C. Swanson, and J. R. Turner, "Alcohol, Intestinal Bacterial Growth, Intestinal Permeability to Endotoxin, and Medical Consequences: Summary of a Symposium," *Alcohol* 42, no. 5 (Aug. 2008): 349–61.

42 A. Keshavarzian, S. Choudhary, E. W. Holmes, S. Yong, A. Banan, S. Jakate, and J. Z. Fields, "Preventing Gut Leakiness by Oats Supplementation Ameliorates Alcohol-induced Liver Damage in Rats," *Journal of Pharmacology and Experimental Therapeutics* 299, no. 2 (Nov. 2001): 442–48.

43 A. Mackie, N. Rigby, P. Harvey, and B. Bajka, "Increasing Dietary Oat Fibre Decreases the Permeability of Intestinal Mucus," *Journal of Functional Foods* 26 (Oct. 2016): 418–27.

44 G. R. Swanson, A. Gorenz, M. Shaikh, V. Desai, C. Forsyth, L. Fogg, H. J. Burgess, and A. Keshavarzian, "Decreased Melatonin Secretion is Associated with Increased Intestinal Permeability and Marker of Endotoxemia in Alcoholics," *American Journal of Physiology-Gastrointestinal and Liver Physiology* 308, no. 12 (Jun. 2015): G1004–11.

45 S. Toden, A. R. Bird, D. L. Topping, and M. A. Conlon, "Resistant Starch Prevents Colonic DNA Damage Induced by High Dietary Cooked Red Meat or Casein in Rats," *Cancer Biology & Therapy* 5, no. 3 (Mar. 2006): 267–72.

46 S. Sonia, F. Witjaksono, and R. Ridwan, "Effect of Cooling of Cooked White Rice on Resistant Starch Content and Glycemic Response," *Asia Pacific Journal of Clinical Nutrition* 24, no. 4 (2015): 620–25.

47 Y. Murakami, S. Tanabe, and T. Suzuki, "High-fat Diet-induced Intestinal Hyperpermeability is Associated with Increased Bile Acids in the Large Intestine of Mice," *Journal of Food Science* 81, no. 1 (Jan. 2016): H216–22.

48 F. Bianchini, G. Caderni, P. Dolara, L. Fantetti, and D. Kriebel, "Effect of Dietary Fat, Starch and Cellulose on Fecal Bile Acids in Mice," *Journal of Nutrition* 119, no. 11 (Nov. 1989): 1617–24.

49 B. Benoit, P. Plaisancié, A. Géloën, M. Estienne, C. Debard, E. Meugnier, E. Loizon, P. Daira, J. Bodennec, O. Cousin, H. Vidal,

F. Laugerette, and M. C. Michalski, "Pasture v. Standard Dairy Cream in High-Fat Diet-Fed Mice: Improved Metabolic Outcomes and Stronger Intestinal Barrier," *British Journal of Nutrition* 112, no. 4 (Aug. 2014): 520–35.

50 J. Suez, T. Korem, D. Zeevi, G. Zilberman-Schapira, C. A. Thaiss, O. Maza, D. Israeli, N. Zmora, S. Gilad, A. Weinberger, Y. Kuperman, A. Harmelin, I. Kolodkin-Gal, H. Shapiro, Z. Halpern, E. Segal, and E. Elinav, "Artificial Sweeteners Induce Glucose Intolerance by Altering the Gut Microbiota," *Nature* 514, no. 7521 (Oct. 2014): 181–86.

51 A. Lerner and T. Matthias, "Changes in Intestinal Tight Junction Permeability Associated with Industrial Food Additives Explain the Rising Incidence of Autoimmune Disease," *Autoimmunity Reviews* 14, no. 6 (Jun. 2015): 479–89.

52 B. Chassaing, O. Koren, J. K. Goodrich, A. C. Poole, S. Srinivasan, R. E. Ley, and A. T. Gewirtz, "Dietary Emulsifiers Impact the Mouse Gut Microbiota Promoting Colitis and Metabolic Syndrome," *Nature* 519, no. 7541 (Mar. 2015): 92–96.

53 N. A. Bokulich and M. J. Blaser, "A Bitter Aftertaste: Unintended Effects of Artificial Sweeteners on the Gut Microbiome," *Cell Metabolism* 20, no. 5 (Nov. 2014): 701–3.

54 B. Chassaing, O. Koren, J. K. Goodrich, A. C. Poole, S. Srinivasan, R. E. Ley, and A. T. Gewirtz, "Dietary Emulsifiers Impact the Mouse Gut Microbiota Promoting Colitis and Metabolic Syndrome," *Nature* 519, no. 7541 (Mar. 2015): 92–96.

55 K. Yoshikawa, C. Kurihara, H. Furuhashi, T. Takajo, K. Maruta, Y. Yasutake, H. Sato, K. Narimatsu, Y. Okada, M. Higashiyama, C. Watanabe, S. Komoto, K. Tomita, S. Nagao, S. Miura, H. Tajiri, and R. Hokari, "Psychological Stress Exacerbates NSAID-induced Small Bowel Injury by Inducing Changes in Intestinal Microbiota and Permeability via Glucocorticoid Receptor Signaling," *Journal of Gastroenterology* 52, no. 1 (Apr. 2016): 61–71.

56 Mackie et al., "Increasing Dietary Oat Fibre Decreases the Permeability of Intestinal Mucus," 418–27.

57 A. R. Mackie, A. Macierzanka, K. Aarak, N. M. Rigby, R. Parker, G. A. Channell, S. E. Harding, and B. H. Bajka, "Sodium Alginate Decreases the Permeability of Intestinal Mucus," *Food Hydrocolloids* 52 (Jan. 2016): 749–55.

58 B. Wang, G. Wu, Z. Zhou, Z. Dai, Y. Sun, Y. Ji, W. Li, W. Wang, C. Liu, F. Han, and Z. Wu, "Glutamine and Intestinal Barrier Function," *Amino Acids* 47, no. 10 (Oct. 2015): 2143–54.

59   C. M. Lenders, S. Liu, D. W. Wilmore, L. Sampson, L. W. Dough-
     erty, D. Spiegelman, and W. C. Willett, "Evaluation of a Novel Food
     Composition Database That Includes Glutamine and Other Amino
     Acids Derived from Gene Sequencing Data," *European Journal of
     Clinical Nutrition* 63, no. 12 (Dec. 2009): 1433–39.

60   G. C. Sturniolo, V. Di Leo, A. Ferronato, A. D'Odorico, and R. D'Inca,
     "Zinc Supplementation Tightens 'Leaky Gut' in Crohn's Disease," *In-
     flammatory Bowel Diseases* 7, no. 2 (2001): 94–98.

61   G. C. Sturniolo, W. Fries, E. Mazzon, V. Di Leo, M. Barollo, and
     R. D'Inca, "Effect of Zinc Supplementation on Intestinal Permea-
     bility in Experimental Colitis," *Journal of Laboratory and Clinical
     Medicine* 139, no. 5 (2002): 311–15.

62   E. V. Boll, L. M. Ekström, C. M. Courtin, J. A. Delcour, A. C. Nils-
     son, I. M. Björck, and E. M. Östman, "Effects of Wheat Bran Ex-
     tract Rich in Arabinoxylan Oligosaccharides and Resistant Starch
     on Overnight Glucose Tolerance and Markers of Gut Fermentation
     in Healthy Young Adults," *European Journal of Nutrition* 55, no. 4
     (Jul. 2015): 1661–70.

63   J. A. Cho and E. Park, "Curcumin Utilizes the Anti-inflammatory
     Response Pathway to Protect the Intestine against Bacterial Inva-
     sion," *Nutrition Research and Practice* 9, no. 2 (Apr. 2015): 117–22.

64   L. Vecchi Brumatti, A. Marcuzzi, P. M. Tricarico, V. Zanin, M. Gi-
     rardelli, and A. M. Bianco, "Curcumin and Inflammatory Bowel
     Disease: Potential and Limits of Innovative Treatments," *Molecules*
     19, no. 12 (Dec. 2014): 21127–53.

65   A. Assa, L. Vong, L. J. Pinnell, N. Avitzur, K. C. Johnson-Henry,
     and P. M. Sherman, "Vitamin D Deficiency Promotes Epithelial
     Barrier Dysfunction and Intestinal Inflammation," *Journal of In-
     fectious Diseases* 210, no. 8 (Oct. 2014): 1296–1305.

66   T. Raftery, A. R. Martineau, C. L. Greiller, S. Ghosh, D. McNamara,
     K. Bennett, J. Meddings, and M. O'Sullivan, "Effects of Vitamin D
     Supplementation on Intestinal Permeability, Cathelicidin and Dis-
     ease Markers in Crohn's Disease: Results from a Randomized Double-
     blind Placebo-controlled Study," *United European Gastroenterology
     Journal* 3, no. 3 (Jun. 2015): 294–302.

67   M. Bashir, B. Prietl, M. Tauschmann, et al., "Effects of High Doses
     of Vitamin D3 on Mucosa-associated Gut Microbiome Vary be-
     tween Regions of the Human Gastrointestinal Tract," *European
     Journal of Nutrition* 55 (2016): 1479–89.

68 A. Reyes, M. Haynes, N. Hanson, F. E. Angly, A. C. Heath, F. Rohwer, and J. I. Gordon, "Viruses in the Faecal Microbiota of Monozygotic Twins and their Mothers," *Nature* 466, no. 7304 (Jul. 2010): 334–38.

69 http://www.nature.com/news/2010/100714/full/news.2010.353 .html. Retrieved Jul. 15, 2016.

70 L. A. David, C. F. Maurice, R. N. Carmody, et al., "Diet Rapidly and Reproducibly Alters the Human Gut Microbiome," *Nature* 505 (2014): 559–63.

71 X. Yang, E. Twitchell, G. Li, K. Wen, M. Weiss, J. Kocher, S. Lei, A. Ramesh, E. P. Ryan, and L. Yuan, "High Protective Efficacy of Rice Bran against Human Rotavirus Diarrhea via Enhancing Probiotic Growth, Gut Barrier Function, and Innate Immunity," *Scientific Reports* 5 (Oct. 2015): 15004.

72 I. Martínez, J. M. Lattimer, K. L. Hubach, J. A. Case, J. Yang, C. G. Weber, J. A. Louk, D. J. Rose, G. Kyureghian, D. A. Peterson, M. D. Haub, and J. Walter, "Gut Microbiome Composition is Linked to Whole Grain-induced Immunological Improvements," *ISME Journal* 7, no. 2 (Feb. 2013): 269–80.

73 A. Basson, A. Trotter, A. Rodriguez-Palacios, and F. Cominelli, "Mucosal Interactions between Genetics, Diet, and Microbiome in Inflammatory Bowel Disease," *Frontiers in Immunology* 7 (Aug. 2016): 290.

74 G. Winther, B. M. Pyndt Jørgensen, B. Elfving, D. S. Nielsen, P. Kihl, S. Lund, D. B. Sørensen, and G. Wegener, "Dietary Magnesium Deficiency Alters Gut Microbiota and Leads to Depressive-like Behaviour," *Acta Neuropsychiatrica* 27, no. 3 (Feb. 2015): 1–9.

75 K. A. Whalen, M. L. McCullough, W. D. Flanders, T. J. Hartman, S. Judd, and R. M. Bostick, "Paleolithic and Mediterranean Diet Pattern Scores Are Inversely Associated with Biomarkers of Inflammation and Oxidative Balance in Adults," *Journal of Nutrition* 146, no. 6 (Jun. 2016): 1217–26.

76 C. Rendeiro, D. Vauzour, R. J. Kean, L. T. Butler, M. Rattray, J. P. Spencer, and C. M. Williams, "Blueberry Supplementation Induces Spatial Memory Improvements and Region-specific Regulation of Hippocampal BDNF mRNA Expression in Young Rats," *Psychopharmacology* 223, no. 3 (Oct. 2012): 319–30.

77 R. Krikorian, M. D. Shidler, T. A. Nash, W. Kalt, M. R. Vinqvist-Tymchuk, B. Shukitt-Hale, and J. A. Joseph, "Blueberry Supplementation

Improves Memory in Older Adults," *Journal of Agricultural and Food Chemistry* 58 (2010): 3996–4000.

78   E. E. Devore, J. H. Kang, M. M. Breteler, and F. Grodstein, "Dietary Intakes of Berries and Flavonoids in Relation to Cognitive Decline," *Annals of Neurology* 72, no. 1 (Jul. 2012): 135–43.

79   https://www.sciencedaily.com/releases/2016/03/160314084821.htm.

80   D. Kelly, R. F. Coen, K. O. Akuffo, S. Beatty, J. Dennison, R. Moran, J. Stack, A. N. Howard, R. Mulcahy, and J. M. Nolan, "Cognitive Function and Its Relationship with Macular Pigment Optical Density and Serum Concentrations of its Constituent Carotenoids," *Journal of Alzheimer's Disease* 48, no. 1 (2015): 261–77.

81   Y. Ozawa, N. Nagai, M. Suzuki, T. Kurihara, H. Shinoda, M. Watanabe, and K. Tsubota, "Effects of Constant Intake of Lutein-rich Spinach on Macular Pigment Optical Density: A Pilot Study," *Nippon Ganka Gakkai Zasshi* 120, no. 1 (Jan. 2016): 41–48.

82   M. Govender, Y. E. Choonara, S. van Vuuren, P. Kumar, L. C. du Toit, and V. Pillay, "Design and Evaluation of an Oral Multiparticulate System for Dual Delivery of Amoxicillin and Lactobacillus acidophilus," *Future Microbiology* 11 (Sept. 2016): 1133–45.

83   J. D. Galley and M. T. Bailey, "Impact of Stressor Exposure on the Interplay between Commensal Microbiota and Host Inflammation," *Gut Microbes* 5 (2014): 390–96.

84   M. T. Bailey and C. L. Coe, "Maternal Separation Disrupts the Integrity of the Intestinal Microflora in Infant Rhesus Monkeys," *Developmental Psychobiology* 35 (1999): 146–55.

85   M. Tanida and K. Nagai, "Electrophysiological Analysis of the Mechanism of Autonomic Action by Lactobacilli," *Bioscience and Microflora* 30, no. 4 (2011): 99–109.

86   M. Nishimura, T. Ohkawara, K. Tetsuka, et al., "Effects of Yogurt Containing *Lactobacillus plantarum* HOKKAIDO on Immune Function and Stress Markers," *Journal of Traditional and Complementary Medicine* 6, no. 3 (2016): 275–80.

87   A. V. Rao, A. C. Bested, T. M. Beaulne, M. A. Katzman, C. Iorio, J. M. Berardi, and A. C. Logan, "A Randomized, Double-blind, Placebo-controlled Pilot Study of a Probiotic in Emotional Symptoms of Chronic Fatigue Syndrome," *Gut Pathogens* 1, no. 1 (Mar. 2009): 6.

88   M. Messaoudi, R. Lalonde, N. Violle, H. Javelot, D. Desor, A. Nejdi, J. F. Bisson, C. Rougeot, M. Pichelin, M. Cazaubiel, and J. M. Cazaubiel, "Assessment of Psychotropic-like Properties of a Probiotic Formulation (*Lactobacillus helveticus* R0052 and *Bifidobacterium*

longum R0175) in Rats and Human Subjects," *British Journal of Nutrition* 105 (2011): 755–64.

89 S. R. Knowles, E. A. Nelson, and E. A. Palombo, "Investigating the Role of Perceived Stress on Bacterial Flora Activity and Salivary Cortisol Secretion: A Possible Mechanism Underlying Susceptibility to Illness," *Biological Psychology* 77 (2008): 132–37.

90 L. Desbonnet, L. Garrett, G. Clarke, B. Kiely, J. F. Cryan, and T. G. Dinan, "Effects of the Probiotic *Bifidobacterium infantis* in the Maternal Separation Model of Depression," *Neuroscience* 170 (2010): 1179–88.

91 D. Benton, C. Williams, and A. Brown, "Impact of Consuming a Milk Drink Containing a Probiotic on Mood and Cognition," *European Journal of Clinical Nutrition* 61, no. 3 (Mar. 2007): 355–61.

92 K. Tillisch, J. Labus, L. Kilpatrick, et al., "Consumption of Fermented Milk Product with Probiotic Modulates Brain Activity," *Gastroenterology* 144, no. 7 (2013): 1394–1401.

93 J. A. Bravo, P. Forsythe, M. V. Chew, E. Escaravage, H. M. Savignac, T. G. Dinan, et al., "Ingestion of *Lactobacillus* Strain Regulates Emotional Behavior and Central GABA Receptor Expression in a Mouse via the Vagus Nerve," *Proceedings of the National Academy of Sciences of the United States of America* 108 (2011): 16050–55.

94 P. A. Mackowiak, "Recycling Metchnikoff: Probiotics, the Intestinal Microbiome and the Quest for Long Life," *Frontiers in Public Health* 1 (2013): 52.

95 Z. Xu and R. Knight, "Dietary Effects on Human Gut Microbiome Diversity," *British Journal of Nutrition* 113, suppl. 0 2015): S1–S5.

96 M. Mohamadshahi, M. Veissi, F. Haidari, H. Shahbazian, G. A. Kaydani, and F. Mohammadi, "Effects of Probiotic Yogurt Consumption on Inflammatory Biomarkers in Patients with Type 2 Diabetes," *BioImpacts* 4, no. 2 (2014): 83–88.

97 http://www.scientificamerican.com/article/gut-second-brain/. Retrieved Jul. 26, 2015.

98 T. T. Haug, A. Mykletun, and A. A. Dahl, "Are Anxiety and Depression Related to Gastrointestinal Symptoms in the General Population?" *Scandinavian Journal of Gastroenterology* 37, no. 3 (Mar. 2002): 294–98.

99 R. Spiller, Q. Aziz, F. Creed, A. Emmanuel, L. Houghton, P. Hungin, R. Jones, D. Kumar, G. Rubin, N. Trudgill, and P. Whorwell, "Guidelines on the Irritable Bowel Syndrome: Mechanisms and Practical Management," *Gut* 56, no. 12 (Dec. 2007): 1770–98.

100 R. M. Lovell and A. C. Ford, "Global Prevalence of and Risk Factors for Irritable Bowel Syndrome: A Meta-analysis," *Clinical Gastroenterology and Hepatology* 10, no. 7 (Jul. 2012): 712–21.

101 M. M. Wouters, S. van Wanrooy, A. Nguyen, J. Dooley, J. Aguilera-Lizarraga, W. van Brabant, J. E. Garcia-Perez, L. van Oudenhove, M. van Ranst, J. Verhaegen, A. Liston, and G. Boeckxstaens, "Psychological Comorbidity Increases the Risk for Postinfectious IBS Partly by Enhanced Susceptibility to Develop Infectious Gastroenteritis," *Gut* 65, no. 8 (Aug. 2016): 1279–88, doi: 10.1136/gutjnl-2015-309460.

102 I. O. Olubuyide, F. Olawuyi, and A. A. Fasanmade, "A Study of Irritable Bowel Syndrome Diagnosed by Manning Criteria in an African Population," *Digestive Diseases and Sciences* 40 (1995): 983–85.

103 H. J. Jung, M. I. Park, W. Moon, S. J. Park, H. H. Kim, E. J. Noh, G. J. Lee, J. H. Kim, and D. G. Kim, "Are Food Constituents Relevant to the Irritable Bowel Syndrome in Young Adults? A Rome III based Prevalence Study of the Korean Medical Students," *Journal of Neurogastroenterology and Motility* 17 (2011): 294–99.

104 L. C. Phua, C. H. Wilder-Smith, Y. M. Tan, T. Gopalakrishnan, R. K. Wong, X. Li, M. E. Kan, J. Lu, A. Keshavarzian, and E. C. Chan, "Gastrointestinal Symptoms and Altered Intestinal Permeability Induced by Combat Training Are Associated with Distinct Metabotypic Changes," *Journal of Proteome Research* 14, no. 11 (Nov. 2015): 4734–42.

105 S. Seyedmirzaee, M. M. Hayatbakhsh, B. Ahmadi, N. Baniasadi, A. M. Bagheri Rafsanjani, A. R. Nikpoor, and M. Mohammadi, "Serum Immune Biomarkers in Irritable Bowel Syndrome," *Clinics and Research in Hepatology and Gastroenterology* 40, no. 5 (Nov. 2016): 631–37.

106 Y. T. Lee, L. Y. Hu, C. C. Shen, M. W. Huang, S. J. Tsai, A. C. Yang, C. K. Hu, C. L. Perng, Y. S. Huang, and J. H. Hung, "Risk of Psychiatric Disorders following Irritable Bowel Syndrome: A Nationwide Population-Based Cohort Study," *PLoS ONE* 10, no. 7 (Jul. 2015): e0133283.

107 C. Tana, Y. Umesaki, A. Imaoka, T. Handa, M. Kanazawa, and S. Fukudo, "Altered Profiles of Intestinal Microbiota and Organic Acids May Be the Origin of Symptoms in Irritable Bowel Syndrome," *Neurogastroenterology & Motility* 22, no. 5 (May 2010): 512–19, e114–15.

108 S. L. Eswaran, W. D. Chey, T. Han-Markey, S. Ball, and K. Jackson, "A Randomized Controlled Trial Comparing the Low FODMAP Diet

vs. Modified NICE Guidelines in US Adults with IBS-D," *American Journal of Gastroenterology* 111, no. 12 (Dec. 2016): 1824–32.

109 L. Böhn, S. Störsrud, T. Liljebo, L. Collin, P. Lindfors, H. Törnblom, and M. Simrén, "Diet Low in FODMAPs Reduces Symptoms of Irritable Bowel Syndrome as Well as Traditional Dietary Advice: A Randomized Controlled Trial," *Gastroenterology* 149, no. 6 (Nov. 2015): 1399–1407, doi: 10.1053/j.gastro.2015.07.054.

110 T. N. Hustoft, T. Hausken, S. O. Ystad, J. Valeur, K. Brokstad, J. G. Hatlebakk, and G. A. Lied, "Effects of Varying Dietary Content of Fermentable Short-chain Carbohydrates on Symptoms, Fecal Microenvironment, and Cytokine Profiles in Patients with Irritable Bowel Syndrome," *Neurogastroenterology & Motility* (Oct. 2016) (*epub ahead of print.*) doi: 10.1111/nmo.12969.

111 D. Lis, K. D. Ahuja, T. Stellingwerff, C. M. Kitic, and J. Fell, "Case Study: Utilizing a Low FODMAP Diet to Combat Exercise-induced Gastrointestinal Symptoms," *International Journal of Sport Nutrition and Exercise Metabolism* 26, no. 5 (Oct. 2016): 481–87.

112 D. Lis, K. D. Ahuja, T. Stellingwerff, C. M. Kitic, and J. Fell, "Food Avoidance in Athletes: FODMAP Foods on the List," *Applied Physiology, Nutrition, and Metabolism* 41, no. 9 (Sept. 2016): 1002–4.

113 Y. Junker, S. Zeissig, S. J. Kim, et al., "Wheat Amylase Trypsin Inhibitors Drive Intestinal Inflammation via Activation of Toll-like Receptor 4," *Journal of Experimental Medicine* 209 (2012): 2395–2408.

114 S. Golley, N. Corsini, D. Topping, et al., "Motivations for Avoiding Wheat Consumption in Australia: Results from a Population Survey," *Public Health Nutrition* 18 (2015): 490–9.

115 A. Carroccio, P. Mansueto, G. Iacono, et al., "Non-celiac Wheat Sensitivity Diagnosed by Double-blind Placebo-controlled Challenge: Exploring a New Clinical Entity," *American Journal of Gastroenterology* 107 (2012): 1898–1906.

116 T. Thompson, "Wheat Starch, Gliadin, and the Gluten-free Diet," *Journal of the American Dietetic Association* 101, no. 12 (Dec. 2001): 1456–59.

117 U. Volta, M. I. Pinto-Sanchez, E. Boschetti, G. Caio, R. De Giorgio, and E. F. Verdu, "Dietary Triggers in Irritable Bowel Syndrome: Is There a Role for Gluten?," *Journal of Neurogastroenterology and Motility* 22, no. 4 (Oct. 2016): 547–57.

118 A. P. Marum et al., "A Low Fermentable Oligo-di-mono Saccharides and Polyols (FODMAP) Diet Reduced Pain and Improved

Daily Life in Fibromyalgia Patients," *Scandinavian Journal of Pain* 13 (Oct. 2016): 166–72.

119 R. Nisihara, A. P. Marques, A. Mei, and T. Skare, "Celiac Disease and Fibromyalgia: Is There an Association?," *Revista Espanola de Enfermedades Digestivas* 108, no. 2 (Feb. 2016): 107–8.

120 G. L. Austin, C. B. Dalton, Y. Hu, C. B. Morris, J. Hankins, S. R. Weinland, E. C. Westman, W. S. Yancy, Jr., and D. A. Drossman, "A Very Low-carbohydrate Diet Improves Symptoms and Quality of Life in Diarrhea-predominant Irritable Bowel Syndrome," *Clinical Gastroenterology and Hepatology* 7, no. 6 (Jun. 2009): 706–8.

121 A. Kumar, N. Kumar, J. C. Vij, S. K. Sarin, and B. S. Anand, "Optimum Dosage of Ispaghula Husk in Patients with Irritable Bowel Syndrome: Correlation of Symptom Relief with Whole Gut Transit Time and Stool Weight," *Gut* 28, no. 2 (Feb. 1987): 150–55.

122 K. Lindfors, T. Blomqvist, K. Juuti-Uusitalo, S. Stenman, J. Venäläinen, M. Mäki, and K. Kaukinen, "Live Probiotic Bifidobacterium lactis Bacteria Inhibit the Toxic Effects Induced by Wheat Gliadin in Epithelial Cell Culture," *Clinical & Experimental Immunology* 152, no. 3 (Jun. 2008): 552–58.

123 S. Guandalini, G. Magazzù, A. Chiaro, V. la Balestra, G. Di Nardo, S. Gopalan, A. Sibal, C. Romano, R. B. Canani, P. Lionetti, and M. Setty, "VSL#3 Improves Symptoms in Children with Irritable Bowel Syndrome: A Multicenter, Randomized, Placebo-controlled, Double-blind, Crossover Study," *Journal of Pediatric Gastroenterology and Nutrition* 51, no. 1 (Jul. 2010): 24–30.

124 N. Shivappa, S. E. Steck, T. G. Hurley, J. R. Hussey, and J. R. Hébert, "Designing and Developing a Literature-derived, Population-based Dietary Inflammatory Index," *Public Health Nutrition* 17, no. 8 (Aug. 2014): 1689–96.

125 N. Shivappa, S. E. Steck, T. G. Hurley, J. R. Hussey, Y. Ma, I. S. Ockene, F. Tabung, and J. R. Hébert, "A Population-based Dietary Inflammatory Index Predicts Levels of C-reactive Protein in the Seasonal Variation of Blood Cholesterol Study," *Public Health Nutrition* 17, no. 8 (Aug. 2014): 1825–33.

126 M. Hafizi Abu Bakar, C. Kian Kai, W. N. Wan Hassan, M. R. Sarmidi, H. Yaakob, and H. Zaman Huri, "Mitochondrial Dysfunction as a Central Event for Mechanisms Underlying Insulin Resistance: The Roles of Long Chain Fatty Acids," *Diabetes/Metabolism Research and Reviews* 31, no. 5 (Jul. 2015): 453–75.

127 A. Mancini, E. Imperlini, E. Nigro, C. Montagnese, A. Daniele, S. Orrù, and P. Buono, "Biological and Nutritional Properties of Palm Oil and Palmitic Acid: Effects on Health," *Molecules* 20, no. 9 (Sept. 2015): 17339–61.

128 R. G. Snodgrass, S. Huang, D. Namgaladze, O. Jandali, T. Shao, S. Sama, B. Brüne, and D. H. Hwang, "Docosahexaenoic Acid and Palmitic Acid Reciprocally Modulate Monocyte Activation in Part through Endoplasmic Reticulum Stress," *Journal of Nutritional Biochemistry* 32 (Jun. 2016): 39–45.

129 M. Hafizi et al., "Mitochondrial Dysfunction as a Central Event for Mechanisms Underlying Insulin Resistance: The Roles of Long Chain Fatty Acids," 453–75.

130 S. Montserrat-de la Paz, M. C. Naranjo, B. Bermudez, S. Lopez, W. Moreda, R. Abia, and F. J. Muriana, "Postprandial Dietary Fatty Acids Exert Divergent Inflammatory Responses in Retinal-pigmented Epithelium Cells," *Food & Function* 7, no. 3 (Mar. 2016): 1345–53.

131 L. M. Varela, B. Bermúdez, A. Ortega-Gómez, S. López, R. Sánchez, J. Villar, C. Anguille, F. J. Muriana, P. Roux, and R. Abia, "Postprandial Triglyceride-rich Lipoproteins Promote Invasion of Human Coronary Artery Smooth Muscle Cells in a Fatty-acid Manner through PI3k-Rac1-JNK Signaling," *Molecular Nutrition & Food Research* 58, no. 6 (Jun. 2014): 1349–64.

132 P. D. Cani, R. Bibiloni, C. Knauf, A. Waget, A. M. Neyrinck, N. M. Delzenne, and R. Burcelin, "Changes in Gut Microbiota Control Metabolic Endotoxemia-induced Inflammation in High-Fat Diet-Induced Obesity and Diabetes in Mice," *Diabetes* 57, no. 6 (Jun. 2008): 1470–81.

133 A. L. Kala, V. Joshi, and K. N. Gurudutt, "Effect of Heating Oils and Fats in Containers of Different Materials on Their Trans Fatty Acid Content," *Journal of the Science of Food and Agriculture* 92, no. 11 (Aug. 2012): 2227–33.

134 S. Bhardwaj, S. J. Passi, A. Misra, K. K. Pant, K. Anwar, R. M. Pandey, and V. Kardam, "Effect of Heating/Reheating of Fats/Oils, as Used by Asian Indians, on Trans Fatty Acid Formation," *Food Chemistry* 212 (Dec. 2016): 663–70.

135 S. H. Ley, Q. Sun, W. C. Willett, et al., "Associations between Red Meat Intake and Biomarkers of Inflammation and Glucose Metabolism in Women," *American Journal of Clinical Nutrition* 99, no. 2 (2014): 352–60.

136 F. Alisson-Silva, K. Kawanishi, and A. Varki, "Human Risk of Diseases Associated with Red Meat Intake: Analysis of Current Theories and Proposed Role for Metabolic Incorporation of a Non-human Sialic Acid," *Molecular Aspects of Medicine* 51 (Oct. 2016): 16–30.
137 A. N. Samraj, O. M. Pearce, H. Läubli, A. N. Crittenden, A. K. Bergfeld, K. Banda, C. J. Gregg, A. E. Bingman, P. Secrest, S. L. Diaz, N. M. Varki, and A. Varki, "A Red Meat-derived Glycan Promotes Inflammation and Cancer Progression," *Proceedings of the National Academy of Sciences of the United States of America* 112, no. 2 (Jan. 2015): 542–47.
138 Samraj et al., "A Red Meat-derived Glycan Promotes Inflammation and Cancer Progression," 542–47.
139 F. Alisson-Silva, K. Kawanishi, and A. Varki, "Human Risk of Diseases Associated with Red Meat Intake: Analysis of Current Theories and Proposed Role for Metabolic Incorporation of a Non-human Sialic Acid," *Molecular Aspects of Medicine* 51 (Oct. 2016): 16–30.
130 B. A. 't Hart, "Why Does Multiple Sclerosis only Affect Human Primates?" *Multiple Sclerosis Journal* 22, no. 4 (Apr. 2016): 559–63.
141 P. Eleftheriou, S. Kynigopoulos, A. Giovou, A. Mazmanidi, J. Yovos, P. Skepastianos, E. Vagdatli, C. Petrou, D. Papara, and M. Efterpiou, "Prevalence of Anti-Neu5Gc Antibodies in Patients with Hypothyroidism," *BioMed Research International* 2014 (2014): 963230.
142 T. Norat, S. Bingham, P. Ferrari, et al., "Meat, Fish, and Colorectal Cancer Risk: The European Prospective Investigation into Cancer and Nutrition," *Journal of the National Cancer Institute* 97 (2005): 906–16.
143 L. H. Kushi, C. Doyle, M. McCullough, C. L. Rock, W. Demark-Wahnefried, E. V. Bandera, S. Gapstur, A. V. Patel, K. Andrews, and T. Gansler, "American Cancer Society Guidelines on Nutrition and Physical Activity for Cancer Prevention: Reducing the Risk of Cancer with Healthy Food Choices and Physical Activity," *CA: A Cancer Journal for Clinicians* 62, no. 1 (Jan.–Feb. 2012): 30–67.
144 R. Sinha, N. Rothman, C. P. Salmon, M. G. Knize, E. D. Brown, C. A. Swanson, D. Rhodes, S. Rossi, J. S. Felton, and O. A. Levander, "Heterocyclic Amine Content in Beef Cooked by Different Methods to Varying Degrees of Doneness and Gravy Made from Meat Drippings," *Food and Chemical Toxicology* 36, no. 4 (Apr. 1998): 279–87.
145 K. Puangsombat, W. Jirapakkul, and J. S. Smith, "Inhibitory Activity of Asian Spices on Heterocyclic Amines Formation in Cooked Beef Patties," *Journal of Food Science* 76, no. 8 (Oct. 2011): T174–80.

146 M. Gibis and J. Weiss, "Antioxidant Capacity and Inhibitory Effect of Grape Seed and Rosemary Extract in Marinades on the Formation of Heterocyclic Amines in Fried Beef Patties," *Food Chemistry* 134, no. 2 (Sept. 2012): 766–74.

147 M. Gibis and J. Weiss, "Inhibitory Effect of Marinades with Hibiscus Extract on Formation of Heterocyclic Aromatic Amines and Sensory Quality of Fried Beef Patties," *Meat Science* 85, no. 4 (Aug. 2010): 735–42.

148 A. Nuora, V. S. Chiang, A. M. Milan, M. Tarvainen, S. Pundir, S. Y. Quek, G. C. Smith, J. F. Markworth, M. Ahotupa, D. Cameron-Smith, and K. M. Linderborg, "The Impact of Beef Steak Thermal Processing on Lipid Oxidation and Postprandial Inflammation related Responses," *Food Chemistry* 184 (Oct. 2015): 57–64.

149 J. de Vogel, D. S. Jonker-Termont, M. B. Katan, and R. van der Meer, "Natural Chlorophyll but Not Chlorophyllin Prevents Heme-induced Cytotoxic and Hyperproliferative Effects in Rat Colon," *Journal of Nutrition* 135, no. 8 (Aug. 2005): 1995–2000.

150 T. Norat, A. Lukanova, P. Ferrari, and E. Riboli, "Meat Consumption and Colorectal Cancer Risk: Dose-response Meta-analysis of Epidemiological Studies," *International Journal of Cancer* 98 (2002): 241–56.

151 J. Uribarri, W. Cai, M. Ramdas, S. Goodman, R. Pyzik, X. Chen, L. Zhu, G. E. Striker, and H. Vlassara, "Restriction of Advanced Glycation End Products Improves Insulin Resistance in Human Type 2 Diabetes: Potential Role of AGER1 and SIRT1," *Diabetes Care* 34 (2011): 1610–16.

152 M. Negrean, A. Stirban, B. Stratmann, T. Gawlowski, T. Horstmann, C. Gotting, K. Kleesiek, M. Mueller-Roesel, T. Koschinsky, J. Uri-barri, H. Vlassara, and D. Tschoepe, "Effects of Low- and High-advanced Glycation End Product Meals on Macro- and Microvascular Endothelial Function and Oxidative Stress in Patients with Type 2 Diabetes mellitus," *American Journal of Clinical Nutrition* 85 (2007): 1236–43.

153 G. Zhang, G. Huang, L. Xiao, and A. E. Mitchell, "Determination of Advanced Glycation End Products by LC-MS/MS in Raw and Roasted Almonds (*Prunus dulcis*)," *Journal of Agricultural and Food Chemistry* 59 (2011): 12037–46.

154 J. Uribarri, S. Woodruff, S. Goodman, W. Cai, X. Chen, R. Pyzik, A. Yong, G. E. Striker, and H. Vlassara, "Advanced Glycation End Products in Foods and a Practical Guide to Their Reduction in the Diet," *Journal of the American Dietetic Association* 110 (2010): 911–16.

155 L. Pruimboom, C. L. Raison, and F. A. Muskiet, "Physical Activity Protects the Human Brain against Metabolic Stress Induced by a Postprandial and Chronic Inflammation," *Behavioural Neurology* 2015: 569869.

156 M. A. Faris, S. Kacimi, R. A. Al-Kurd, M. A. Fararjeh, Y. K. Bustanji, M. K. Mohammad, and M. L. Salem, "Intermittent Fasting during Ramadan Attenuates Proinflammatory Cytokines and Immune Cells in Healthy Subjects," *Nutrition Research* 32, no. 12 (Dec. 2012): 947–55.

157 T. Moro, G. Tinsley, A. Bianco, G. Marcolin, Q. F. Pacelli, G. Battaglia, A. Palma, P. Gentil, M. Neri, and A. Paoli, "Effects of Eight Weeks of Time-restricted Feeding (16/8) on Basal Metabolism, Maximal Strength, Body Composition, Inflammation, and Cardiovascular Risk Factors in Resistance-trained Males," *Journal of Translational Medicine* 14, no. 1 (Oct. 2016): 290.

158 J. L. Mauriz, P. S. Collado, C. Veneroso, R. J. Reiter, and J. González-Gallego, "A Review of the Molecular Aspects of Melatonin's Anti-inflammatory Actions: Recent Insights and New Perspectives," *Journal of Pineal Research* 54, no. 1 (2013): 1–14.

159 L. Fontana, "Neuroendocrine Factors in the Regulation of Inflammation: Excessive Adiposity and Calorie Restriction," *Experimental Gerontology* 44 (2009): 41–45.

160 S. D. Katewa, K. Akagi, N. Bose, K. Rakshit, T. Camarella, X. Zheng, D. Hall, S. Davis, C. S. Nelson, R. B. Brem, A. Ramanathan, A. Sehgal, J. M. Giebultowicz, and P. Kapahi, "Peripheral Circadian Clocks Mediate Dietary Restriction-Dependent Changes in Lifespan and Fat Metabolism in Drosophila," *Cell Metabolism* 23, no. 1 (Jan. 2016): 143–54.

161 F. Molica, S. Morel, B. R. Kwak, F. Rohner-Jeanrenaud, and S. Steffens, "Adipokines at the Crossroad between Obesity and Cardiovascular Disease," *Thrombosis and Haemostasis* 113, no. 3 (Mar. 2014): 553–66.

162 T. Karrasch and A. Schaeffler, "Adipokines and the Role of Visceral Adipose Tissue in Inflammatory Bowel Disease," *Annals of Gastroenterology* 29, no. 4 (Oct.–Dec. 2016): 424–38.

163 I. Schlecht, B. Fischer, G. Behrens, and M. F. Leitzmann, "Relations of Visceral and Abdominal Subcutaneous Adipose Tissue, Body Mass Index, and Waist Circumference to Serum Concentrations of Parameters of Chronic Inflammation," *Obesity Facts* 9, no. 3 (2016): 144–57.

164 A. Gummesson, L. M. Carlsson, L. H. Storlien, F. Bäckhed, P. Lundin, L. Löfgren, K. Stenlöf, Y. Y. Lam, B. Fagerberg, and B. Carlsson, "Intestinal Permeability Is Associated with Visceral Adiposity in Healthy Women," *Obesity* 19, no. 11 (Nov. 2011): 2280–82.

165 G. Escobedo, E. López-Ortiz, and I. Torres-Castro, "Gut Microbiota as a Key Player in Triggering Obesity, Systemic Inflammation and Insulin Resistance," *Revista de Investigación Clínica* 66, no. 5 (Sept.–Oct. 2014): 450–59.

166 Karrasch et al., "Adipokines and the Role of Visceral Adipose Tissue in Inflammatory Bowel Disease," 424–38.

167 H. Li, C. Lelliott, P. Håkansson, et al., "Intestinal, Adipose, and Liver Inflammation in Diet-induced Obese Mice," *Metabolism: Clinical and Experimental* 57 (2008): 1704–10.

168 M. Kawano, M. Miyoshi, A. Ogawa, F. Sakai, and Y. Kadooka, "*Lactobacillus gasseri* SBT2055 Inhibits Adipose Tissue Inflammation and Intestinal Permeability in Mice Fed a High-fat Diet," *Journal of Nutritional Science* 5 (2016): e23, doi: 10.1017/jns.2016.12.

169 J. P. Després and I. Lemieux, "Abdominal Obesity and Metabolic Syndrome," *Nature* 444, no. 7121 (Dec. 2006): 881–87.

170 L. E. Gyllenhammer, M. J. Weigensberg, D. Spruijt-Metz, H. Allayee, M. I. Goran, and J. N. Davis, "Modifying Influence of Dietary Sugar in the Relationship between Cortisol and Visceral Adipose Tissue in Minority Youth," *Obesity* 22, no. 2 (2014): 474–81.

171 H. R. Hong, J. O. Jeong, J. Y. Kong, S. H. Lee, S. H. Yang, C. D. Ha, and H. S. Kang, "Effect of Walking Exercise on Abdominal Fat, Insulin Resistance and Serum Cytokines in Obese Women," *Journal of Exercise Nutrition & Biochemistry* 18, no. 3 (Sept. 2014): 277–85.

172 R. Ross, R. Hudson, P. J. Stotz, and M. Lam, "Effects of Exercise Amount and Intensity on Abdominal Obesity and Glucose Tolerance in Obese Adults: A Randomized Trial," *Annals of Internal Medicine* 162, no. 5 (Mar. 2015): 325–34.

173 S. E. Keating, D. A. Hackett, H. M. Parker, H. T. O'Connor, J. A. Gerofi, A. Sainsbury, M. K. Baker, V. H. Chuter, I. D. Caterson, J. George, and N. A. Johnson, "Effect of Aerobic Exercise Training Dose on Liver Fat and Visceral Adiposity," *Journal of Hepatology* 63, no. 1 (Jul. 2015): 174–82.

174 G. R. Logan, N. Harris, S. Duncan, L. D. Plank, F. Merien, and G. Schofield, "Low-active Male Adolescents: A Dose Response to High-intensity Interval Training," *Medicine & Science in Sports & Exercise* 48, no. 3 (Mar. 2016): 481–90.

175 K. E. Koopman, M. W. Caan, A. J. Nederveen, et al., "Hypercaloric Diets with Increased Meal Frequency, but not Meal Size, Increase Intrahepatic Triglycerides: A Randomized Controlled Trial," *Hepatology* 60, no. 2 (2014): 545–53.

176 M. Gleeson, N. C. Bishop, D. J. Stensel, et al., "The Anti-inflammatory Effects of Exercise: Mechanisms and Implications for the Prevention and Treatment of Disease," *Nature Reviews Immunology* 11 (2011): 607–15.

177 G. I. Lancaster and M. A. Febbraio, "The Immunomodulating Role of Exercise in Metabolic Disease," *Trends in Immunology* (2014): 262–69.

178 C. Kasapis and P. D. Thompson, "The Effects of Physical Activity on Serum C-reactive Protein and Inflammatory Markers: A Systematic Review," *Journal of the American College of Cardiology* 45 (2005): 1563–69.

179 A. Masuda, M. Nakazato, T. Kihara, S. Minagoe, and C. Tei, "Repeated Thermal Therapy Diminishes Appetite Loss and Subjective Complaints in Mildly Depressed Patients," *Psychosomatic Medicine* 67, no. 4 (Jul.–Aug. 2005): 643–47.

180 C. W. Janssen, C. A. Lowry, M. R. Mehl, J. J. Allen, K. L. Kelly, D. E. Gartner, A. Medrano, T. K. Begay, K. Rentscher, J. J. White, A. Fridman, L. J. Roberts, M. L. Robbins, K. U. Hanusch, S. P. Cole, and C. L. Raison, "Whole-body Hyperthermia for the Treatment of Major Depressive Disorder: A Randomized Clinical Trial," *JAMA Psychiatry* 73, no. 8 (Aug. 2016): 789–95.

181 M. Bauer, S. Berman, T. Stamm, M. Plotkin, M. Adli, M. Pilhatsch, E. D. London, G. S. Hellemann, P. C. Whybrow, and F. Schlagenhauf, "Levothyroxine Effects on Depressive Symptoms and Limbic Glucose Metabolism in Bipolar Disorder: A Randomized, Placebo-controlled Positron Emission Tomography Study," *Molecular Psychiatry* 21, no. 2 (Feb. 2016): 229–36.

182 M. Krause, M. S. Ludwig, T. G. Heck, and H. K. Takahashi, "Heat Shock Proteins and Heat Therapy for Type 2 Diabetes: Pros and Cons," *Current Opinion in Clinical Nutrition and Metabolic Care* 18, no. 4 (Jul. 2015): 374–80.

183 T. Laukkanen, S. Kunutsor, J. Kauhanen, and J. A. Laukkanen, "Sauna Bathing is Inversely Associated with Dementia and Alzheimer's Disease in Middle-aged Finnish Men," *Age and Ageing* 1, no. 5 (Dec. 2016) (*epub ahead of print*) doi:10.1093/ageing/afw212.

184 P. Cassano, S. R. Petrie, M. R. Hamblin, T. A. Henderson, and D. V. Iosifescu, "Review of Transcranial Photobiomodulation for Major Depressive Disorder: Targeting Brain Metabolism, Inflammation, Oxidative Stress, and Neurogenesis," *Neurophotonics* 3, no. 3 (Jul. 2016): 031404.

185 L. Detaboada et al., "Transcranial Application of Low-energy Laser Irradiation Improves Neurological Deficits in Rats Following Acute Stroke," *Lasers in Surgery and Medicine* 38, no. 1 (2006): 70–73.

186 H. Araki et al., "Reduction of Interleukin-6 Expression in Human Synoviocytes and Rheumatoid Arthritis Rat Joints by Linear Polarized Near Infrared Light (Superlizer) Irradiation," *Laser Therapy* 20, no. 4 (2011): 293.

187 M. Yamaura et al., "Low Level Light Effects on Inflammatory Cytokine Production by Rheumatoid Arthritis Synoviocytes," *Lasers in Surgery and Medicine* 41, no. 4 (2009): 282–90.

188 J. Khuman et al., "Low-level Laser Light Therapy Improves Cognitive Deficits and Inhibits Microglial Activation after Controlled Cortical Impact in Mice," *Journal of Neurotrauma* 29, no. 2 (2012): 408–17.

189 J. C. Rojas, A. K. Bruchey, and F. Gonzalez-Lima, "Low-level Light Therapy Improves Cortical Metabolic Capacity and Memory Retention," *Journal of Alzheimer's Disease* 32, no. 3 (2012): 741–52.

190 D. Barrett and F. Gonzalez-Lima, "Transcranial Infrared Laser Stimulation Produces Beneficial Cognitive and Emotional Effects in Humans," *Neuroscience* 230 (2013): 13–23.

191 X. Qin and E. A. Deitch, "Dissolution of Lipids from Mucus: A Possible Mechanism for Prompt Disruption of Gut Barrier Function by Alcohol," *Toxicology Letters* 232, no. 2 (Jan. 2015): 356–62.

192 J. Connor, "Alcohol Consumption as a Cause of Cancer," *Addiction* 112, no. 2 (Feb. 2017): 222–28.

CHAPTER 7

1 R. Nirupama, M. Devaki, and H. N. Yajurvedi, "Chronic Stress and Carbohydrate Metabolism: Persistent Changes and Slow Return to Normalcy in Male Albino Rats," *Stress* 15, no. 3 (May 2012): 262–71.

2 S. Garbarino and N. Magnavita, "Work Stress and Metabolic Syndrome in Police Officers. A Prospective Study," ed. V. Grolmusz, *PLoS ONE* 10, no. 12 (2015): e0144318.

3   A. Hino, A. Inoue, K. Mafune, T. Nakagawa, T. Hayashi, and H. Hiro, "Changes in the Psychosocial Work Characteristics and Insulin Resistance among Japanese Male Workers: A Three-year Follow-up Study," *Journal of Occupational Health* 58, no. 6 (Nov. 2016): 543–62.

4   B. Schmidt, J. A. Bosch, M. N. Jarczok, R. M. Herr, A. Loerbroks, A. E. van Vianen, and J. E. Fischer, "Effort-reward Imbalance Is Associated with the Metabolic Syndrome—Findings from the Mannheim Industrial Cohort Study (MICS)," *International Journal of Cardiology* 178 (Jan. 2015): 24–28.

5   T. Almadi, I. Cathers, and C. M. Chow, "Associations among Work-related Stress, Cortisol, Inflammation, and Metabolic Syndrome," *Psychophysiology* 50, no. 9 (Sept. 2013): 821–30.

6   V. K. Tsenkova, D. Carr, C. L. Coe, and C. D. Ryff, "Anger, Adiposity, and Glucose Control in Nondiabetic Adults: Findings from Midus II," *Journal of Behavioral Medicine* 37, no. 1 (2014): 37–46.

7   K. Räikkönen, L. Keltikangas-Järvinen, and A. Hautanen, "The Role of Psychological Coronary Risk Factors in Insulin and Glucose Metabolism," *Journal of Psychosomatic Research* 38, no. 7 (Oct. 1994): 705–13.

8   S. H. Boyle, A. Georgiades, B. H. Brummett, et al., "Associations between Central Nervous System Serotonin, Fasting Glucose and Hostility in African American Females," *Annals of Behavioral Medicine: A Publication of the Society of Behavioral Medicine* 49, no. 1 (2015): 49–57.

9   A. Quincozes-Santos, L. D. Bobermin, A. M. de Assis, C. A. Gonçalves, and D. O. Souza, "Fluctuations in Glucose Levels Induce Glial Toxicity with Glutamatergic, Oxidative and Inflammatory Implications," *Biochimica et Biophysica Acta* 1863, no. 1 (Jan. 2017): 1–14.

10   D. E. Rivera-Aponte, M. P. Méndez-González, A. F. Rivera-Pagán, Y. V. Kucheryavykh, L. Y. Kucheryavykh, S. N. Skatchkov, and M. J. Eaton, "Hyperglycemia Reduces Functional Expression of Astrocytic Kir4.1 channels and Glial Glutamate Uptake," *Neuroscience* 310 (Dec. 2015): 216–23.

11   When one brain cell passes a message to another and wants the recipient to become activated or "excited," it sends it the chemical messenger **glutamate**. Glutamate will keep exciting the recipient brain until it is cleared away. If it is not cleared away, the brain cell can die of too much excitation or "excitotoxicity." Elevated blood sugar, at levels comparable to those seen in the setting of insulin resistance and diabetes, can interfere with the clearing away of glutamate.

12   G. B. Stefano, S. Challenger, and R. M. Kream, "Hyperglycemia-associated Alterations in Cellular Signaling and Dysregulated Mitochondrial Bioenergetics in Human Metabolic Disorders," *European Journal of Nutrition* 55, no. 8 (Dec. 2016): 2339–2345.

13   W. Cai, J. Uribarri, L. Zhu, X. Chen, S. Swamy, Z. Zhao, F. Grosjean, C. Simonaro, G. A. Kuchel, M. Schnaider-Beeri, M. Woodward, G. E. Striker, and H. Vlassara, "Oral Glycotoxins Are a Modifiable Cause of Dementia and the Metabolic Syndrome in Mice and Humans," *Proceedings of the National Academy of Sciences of the United States of America* 111, no. 13 (Apr. 2014): 4940–45.

14   K. A. Page, A. Williamson, N. Yu, E. C. McNay, J. Dzuira, R. J. McCrimmon, and R. S. Sherwin, "Medium-chain Fatty Acids Improve Cognitive Function in Intensively Treated Type 1 Diabetic Patients and Support in Vitro Synaptic Transmission during Acute Hypoglycemia," *Diabetes* 58, no. 5 (May 2009): 1237–44.

15   E. Bullmore and O. Sporns, "The Economy of Brain Network Organization." *Nature Reviews Neuroscience* 13 (2012): 336–49.

16   K. Ishibashi, K. Wagatsuma, K. Ishiwata, and K. Ishii, "Alteration of the Regional Cerebral Glucose Metabolism in Healthy Subjects by Glucose Loading," *Human Brain Mapping* 37, no. 8 (Aug. 2016): 2823–32.

17   M. Brendel, V. Reinisch, E. Kalinowski, J. Levin, A. Delker, S. Därr, O. Pogarell, S. Förster, P. Bartenstein, and A. Rominger, "Hypometabolism in Brain of Cognitively Normal Patients with Depressive Symptoms is Accompanied by Atrophy-Related Partial Volume Effects," *Current Alzheimer Research* 13, no. 5 (2016): 475–86.

18   C. M. Marano, C. I. Workman, C. H. Lyman, E. Kramer, C. R. Hermann, Y. Ma, V. Dhawan, T. Chaly, D. Eidelberg, and G. S. Smith, "The Relationship between Fasting Serum Glucose and Cerebral Glucose Metabolism in Late-life Depression and Normal Aging," *Psychiatry Research* 222, nos. 1–2 (Apr. 2014): 84–90.

19   C. A. Castellano, J. P. Baillargeon, S. Nugent, S. Tremblay, M. Fortier, H. Imbeault, J. Duval, and S. C. Cunnane, "Regional Brain Glucose Hypometabolism in Young Women with Polycystic Ovary Syndrome: Possible Link to Mild Insulin Resistance," *PLoS ONE* 10, no. 12 (Dec. 2015): e0144116.

20   M. Moosavi, N. Naghdi, N. Maghsoudi, and A. S. Zahedi, "The Effect of Intrahippocampal Insulin Microinjection on Spatial Learning and Memory," *Hormones and Behavior* 50 (2006): 748–52.

21 D. R. Marks, K. Tucker, M. A. Cavallin, T. G. Mast, and D. A. Fadool, "Awake Intranasal Insulin Delivery Modifies Protein Complexes and Alters Memory, Anxiety, and Olfactory Behaviors," *Journal of Neuroscience* 29 (2009): 6734–51.

22 C. Benedict, M. Hallschmid, K. Schmitz, B. Schultes, F. Ratter, H. L. Fehm, J. Born, and W. Kern, "Intranasal Insulin Improves Memory in Humans: Superiority of Insulin Aspart," *Neuropsychopharmacology* 32 (2007): 239–43.

23 D. Jakubowicz, J. Wainstein, B. Ahrén, Y. Bar-Dayan, Z. Landau, H. R. Rabinovitz, and O. Froy, "High-energy Breakfast with Low-energy Dinner Decreases Overall Daily Hyperglycaemia in Type 2 Diabetic Patients: A Randomised Clinical Trial," *Diabetologia* 58, no. 5 (May 2015): 912–19.

24 D. Jakubowicz, J. Wainstein, B. Ahren, Z. Landau, Y. Bar-Dayan, and O. Froy, "Fasting until Noon Triggers Increased Postprandial Hyperglycemia and Impaired Insulin Response after Lunch and Dinner in Individuals with Type 2 Diabetes: A Randomized Clinical Trial," *Diabetes Care* 38, no. 10 (Oct. 2015): 1820–26.

25 T. Remer and F. Manz, "Potential Renal Acid Load of Foods and Its Influence on Urine pH," *Journal of the American Dietetic Association* 95, no. 7 (Jul. 1995): 791–97.

26 R. S. Williams, L. K. Heilbronn, D. L. Chen, A. C. Coster, J. R. Greenfield, and D. Samocha-Bonet, "Dietary Acid Load, Metabolic Acidosis and Insulin Resistance—Lessons from Cross-sectional and Overfeeding Studies in Humans," *Clinical Nutrition* 35, no. 5 (Oct. 2016): 1084–90.

27 J. Koska, M. K. Ozias, J. Deer, J. Kurtz, A. D. Salbe, S. M. Harman, and P. D. Reaven, "A Human Model of Dietary Saturated Fatty Acid Induced Insulin Resistance," *Metabolism* 65, no. 11 (Nov. 2016): 1621–28.

28 P. Kiilerich, L. S. Myrmel, E. Fjære, Q. Hao, F. Hugenholtz, S. B. Sonne, M. Derrien, L. M. Pedersen, R. K. Petersen, A. Mortensen, T. R. Licht, M. U. Rømer, U. B. Vogel, L. J. Waagbø, N. Giallourou, Q. Feng, L. Xiao, C. Liu, B. Liaset, M. Kleerebezem, J. Wang, L. Madsen, and K. Kristiansen, "Effect of a Long-term High-protein Diet on Survival, Obesity Development, and Gut Microbiota in Mice," *American Journal of Physiology—Endocrinology and Metabolism* 310, no. 11 (Jun. 2016): E886–99.

29 L. K. Stenman, R. Holma, A. Eggert, and R. Korpela, "A Novel Mechanism for Gut Barrier Dysfunction by Dietary Fat:

Epithelial Disruption by Hydrophobic Bile Acids," *American Journal of Physiology—Gastrointestinal and Liver Physiology* 304, no. 3 (Feb. 2013): G227–34.

30 V. Costarelli and T. A. Sanders, "Acute Effects of Dietary Fat Composition on Postprandial Plasma Bile Acid and Cholecystokinin Concentrations in Healthy Premenopausal Women," *British Journal of Nutrition* 86, no. 4 (Oct. 2001): 471–77.

31 C. Ferreira-Pêgo, N. Babio, M. Bes-Rastrollo, D. Corella, R. Estruch, E. Ros, M. Fitó, L. Serra-Majem, F. Arós, M. Fiol, J. M. Santos-Lozano, C. Muñoz-Bravo, X. Pintó, M. Ruiz-Canela, and J. Salas-Salvadó, "Frequent Consumption of Sugar- and Artificially Sweetened Beverages and Natural and Bottled Fruit Juices Is Associated with an Increased Risk of Metabolic Syndrome in a Mediterranean Population at High Cardiovascular Disease Risk," *Journal of Nutrition* 146, no. 8 (Aug. 2016): 1528–36.

32 M. S. Kim, S. A. Krawczyk, L. Doridot, A. J. Fowler, J. X. Wang, S. A. Trauger, H. L. Noh, H. J. Kang, J. K. Meissen, M. Blatnik, J. K. Kim, M. Lai, and M. A. Herman, "ChREBP Regulates Fructose-induced Glucose Production Independently of Insulin Signaling," *Journal of Clinical Investigation* 126, no. 11 (Sept. 2016): 4372–86.

33 S. E. la Fleur, M. C. Luijendijk, A. J. van Rozen, A. Kalsbeek, and R. A. Adan, "A Free-choice High-fat High-sugar Diet Induces Glucose Intolerance and Insulin Unresponsiveness to a Glucose Load Not Explained by Obesity," *International Journal of Obesity* 35, no. 4 (Apr. 2011): 595–604.

34 S. Lindeberg, M. Eliasson, B. Lindahl, and B. Ahrén, "Low Serum Insulin in Traditional Pacific Islanders—the Kitava Study," *Metabolism* 48, no. 10 (Oct. 1999): 1216–19.

35 S. Geng, W. Zhu, C. Xie, X. Li, J. Wu, Z. Liang, W. Xie, J. Zhu, C. Huang, M. Zhu, R. Wu, and C. Zhong, "Medium-chain Triglyceride Ameliorates Insulin Resistance and Inflammation in High Fat Diet-induced Obese Mice," *European Journal of Nutrition* 55, no. 3 (Apr. 2016): 931–40.

36 M. Sakurai, K. Nakamura, K. Miura, T. Takamura, K. Yoshita, S. Y. Nagasawa, Y. Morikawa, M. Ishizaki, T. Kido, Y. Naruse, M. Nakashima, K. Nogawa, Y. Suwazono, S. Sasaki, and H. Nakagawa, "Dietary Carbohydrate Intake, Presence of Obesity and the Incident Risk of Type 2 Diabetes in Japanese Men," *Journal of Diabetes Investigation* 7, no. 3 (May 2016): 343–51.

37 P. J. Lin and K. T. Borer, "Third Exposure to a Reduced Carbohydrate Meal Lowers Evening Postprandial Insulin and GIP Responses and HOMA-IR Estimate of Insulin Resistance," *PLoS ONE* 11, no. 10 (Oct. 2016): e0165378.

38 R. Salvia, S. D'Amore, G. Graziano, C. Capobianco, M. Sangineto, D. Paparella, P. de Bonfils, G. Palasciano, and M. Vacca, "Short-term Benefits of an Unrestricted-calorie Traditional Mediterranean Diet, Modified with a Reduced Consumption of Carbohydrates at Evening, in Overweight-obese Patients," *International Journal of Food Sciences and Nutrition* 68, no. 2 (Mar. 2017): 234–248.

39 C. Eelderink, M. W. Noort, N. Sozer, M. Koehorst, J. J. Holst, C. F. Deacon, J. F. Rehfeld, K. Poutanen, R. J. Vonk, L. Oudhuis, and M. G. Priebe, "The Structure of Wheat Bread Influences the Postprandial Metabolic Response in Healthy Men," *Food & Function* 6, no. 10 (Oct. 2015): 3236–48.

40 K. S. Juntunen, D. E. Laaksonen, K. Autio, L. K. Niskanen, J. J. Holst, K. E. Savolainen, K. H. Liukkonen, K. S. Poutanen, and H. M. Mykkänen, "Structural Differences between Rye and Wheat Breads but Not Total Fiber Content May Explain the Lower Postprandial Insulin Response to Rye Bread," *American Journal of Clinical Nutrition* 78, no. 5 (Nov. 2003): 957–64.

41 S. Sonia, F. Witjaksono, and R. Ridwan, "Effect of Cooling of Cooked White Rice on Resistant Starch Content and Glycemic Response," *Asia Pacific Journal of Clinical Nutrition* 24, no. 4 (2015): 620–25.

42 Joint WHO/FAO/UNU Expert Consultation, "Protein and Amino Acid Requirements in Human Nutrition," *World Health Organ Technical Report Series* 935 (2007): 1–265, back cover.

43 I. Sluijs, J. W. Beulens, D. L. van der A, A. M. Spijkerman, D. E. Grobbee, and Y. T. van der Schouw, "Dietary Intake of Total, Animal, and Vegetable Protein and Risk of Type 2 Diabetes in the European Prospective Investigation into Cancer and Nutrition (EPIC)-NL Study," *Diabetes Care* 33, no. 1 (Jan. 2010): 43–48.

44 J. Matta, N. Mayo, I. J. Dionne, P. Gaudreau, T. Fulop, D. Tessier, K. Gray-Donald, B. Shatenstein, and J. A. Morais, "Muscle Mass Index and Animal Source of Dietary Protein Are Positively Associated with Insulin Resistance in Participants of the NuAge Study," *Journal of Nutrition Health & Aging* 20, no. 2 (Feb. 2016): 90–97.

45 G. I. Smith, J. Yoshino, K. L. Stromsdorfer, S. J. Klein, F. Magkos, D. N. Reeds, S. Klein, and B. Mittendorfer, "Protein Ingestion

Induces Muscle Insulin Resistance Independent of Leucine-Mediated mTOR Activation," *Diabetes* 64, no. 5 (May 2015): 1555–63.

46 G. I. Smith, J. Yoshino, S. C. Kelly, D. N. Reeds, A. Okunade, B. W. Patterson, S. Klein, and B. Mittendorfer, "High-protein Intake during Weight Loss Therapy Eliminates the Weight-loss-induced Improvement in Insulin Action in Obese Postmenopausal Women," *Cell Reports* 17, no. 3 (Oct. 2016): 849–61.

47 D. Aune, G. Ursin, and M. B. Veierød, "Meat Consumption and the Risk of Type 2 Diabetes: A Systematic Review and Meta-analysis of Cohort Studies," *Diabetologia* 52, no. 11 (Nov. 2009): 2277–87, doi: 10.1007/s00125-009-1481-x.

48 N. Benaicheta, F. Z. Labbaci, M. Bouchenak, and F. O. Boukortt, "Effect of Sardine Proteins on Hyperglycaemia, Hyperlipidaemia and Lecithin: Cholesterol Acyltransferase Activity, in High-Fat Diet-Induced Type 2 Diabetic Rats," *British Journal of Nutrition* 115, no. 1 (Jan. 2016): 6–13.

49 M. S. Ottum and A. M. Mistry, "Advanced Glycation End-products: Modifiable Environmental Factors Profoundly Mediate Insulin Resistance," *Journal of Clinical Biochemistry and Nutrition* 57, no. 1 (Jul. 2015): 1–12.

50 A. Taniguchi-Fukatsu, H. Yamanaka-Okumura, Y. Naniwa-Kuroki, Y. Nishida, H. Yamamoto, Y. Taketani, and E. Takeda, "Natto and Viscous Vegetables in a Japanese-style Breakfast Improved Insulin Sensitivity, Lipid Metabolism and Oxidative Stress in Overweight Subjects with Impaired Glucose Tolerance," *British Journal of Nutrition* 107, no. 8 (Apr. 2012): 1184–91.

51 P. Ebeling, H. Yki-Järvinen, A. Aro, et al., "Glucose and Lipid Metabolism and Insulin Sensitivity in Type 1 Diabetes: The Effect of Guar Gum," *American Journal of Clinical Nutrition* 48, no. 1 (1988): 98–103.

52 C. A. Clark, J. Gardiner, M. I. McBurney, S. Anderson, L. J. Weatherspoon, D. N. Henry, and N. G. Hord, "Effects of Breakfast Meal Composition on Second Meal Metabolic Responses in Adults with Type 2 Diabetes mellitus," *European Journal of Clinical Nutrition* 60, no. 9 (Sept. 2006): 1122–29.

53 M. Drehmer, M. A. Pereira, M. I. Schmidt, B. Del Carmen, M. Molina, S. Alvim, P. A. Lotufo, and B. B. Duncan, "Associations of Dairy Intake with Glycemia and Insulinemia, Independent of Obesity, in Brazilian Adults: The Brazilian Longitudinal Study of Adult

Health (ELSA-Brasil)," *American Journal of Clinical Nutrition* 101, no. 4 (Apr. 2015): 775–82.

54 M. Chen, Q. Sun, E. Giovannucci, D. Mozaffarian, J. E. Manson, W. C. Willett, and F. B. Hu, "Dairy Consumption and Risk of Type 2 Diabetes: 3 Cohorts of US Adults and an Updated Meta-analysis," *BMC Medicine* 12 (Nov. 2014): 215.

55 C. J. Hulston, A. A. Churnside, and M. C. Venables, "Probiotic Supplementation Prevents High-fat, Overfeeding-induced Insulin Resistance in Human Subjects," *British Journal of Nutrition* 113, no. 4 (Feb. 2015): 596–602.

56 Mohamadshahi et al., "Effects of Probiotic Yogurt Consumption on Inflammatory Biomarkers in Patients with Type 2 Diabetes," 83–88.

57 J. Dolpady, C. Sorini, C. Di Pietro, I. Cosorich, R. Ferrarese, D. Saita, M. Clementi, F. Canducci, and M. Falcone, "Oral Probiotic VSL#3 Prevents Autoimmune Diabetes by Modulating Microbiota and Promoting Indoleamine 2,3-Dioxygenase-Enriched Tolerogenic Intestinal Environment," *Journal of Diabetes Research* 2016 (2016): 7569431.

58 Z. Ghorbani, A. Hekmatdoost, and P. Mirmiran, "Anti-hyperglycemic and Insulin Sensitizer Effects of Turmeric and Its Principle [*sic*] Constituent Curcumin," *International Journal of Endocrinology and Metabolism* 12, no. 4 (Oct. 2014): e18081.

59 C. K. Atal, R. K. Dubey, and J. Singh, "Biochemical Basis of Enhanced Drug Bioavailability by Piperine: Evidence That Piperine Is a Potent Inhibitor of Drug Metabolism," *Journal of Pharmacology and Experimental Therapeutics* 232, no. 1 (Jan. 1985): 258–62.

60 G. Shoba, D. Joy, T. Joseph, M. Majeed, R. Rajendran, and P. S. Srinivas, "Influence of Piperine on the Pharmacokinetics of Curcumin in Animals and Human Volunteers," *Planta Medica* 64, no. 4 (May 1998): 353–56.

61 P. Anand, A. B. Kunnumakkara, R. A. Newman, and B. B. Aggarwal, "Bioavailability of Curcumin: Problems and Promises," *Molecular Pharmaceutics* 4, no. 6 (2007): 807–18.

62 S. Prasad, A. K. Tyagi, and B. B. Aggarwal, "Recent Developments in Delivery, Bioavailability, Absorption and Metabolism of Curcumin: The Golden Pigment from Golden Spice," *Cancer Research and Treatment: Official Journal of Korean Cancer Association* 46, no. 1 (2014): 2–18.

63 A. S. Sahib, "Anti-diabetic and Antioxidant Effect of Cinnamon in Poorly Controlled Type-2 Diabetic Iraqi Patients: A Randomized, Placebo-controlled Clinical Trial," *Journal of Intercultural Ethnopharmacology* 5, no. 2 (Feb. 2016): 108–13.

64 J. Hlebowicz, A. Hlebowicz, S. Lindstedt, O. Björgell, P. Höglund, J. J. Holst, G. Darwiche, and L. O. Almér, "Effects of 1 and 3 g Cinnamon on Gastric Emptying, Satiety, and Postprandial Blood Glucose, Insulin, Glucose-dependent Insulinotropic Polypeptide, Glucagon-like Peptide 1, and Ghrelin Concentrations in Healthy Subjects," *American Journal of Clinical Nutrition* 89, no. 3 (Mar. 2009): 815–21.

65 N. Veronese, S. F. Watutantrige, C. Luchini, M. Solmi, G. Sartore, G. Sergi, E. Manzato, M. Barbagallo, S. Maggi, and B. Stubbs, "Effect of Magnesium Supplementation on Glucose Metabolism in People with or at Risk of Diabetes: A Systematic Review and Meta-analysis of Double-blind Randomized Controlled Trials," *European Journal of Clinical Nutrition* 70, no. 12 (Dec. 2016): 1354–59.

66 https://www.cedars-sinai.edu/Patients/Programs-andServices/Documents/CP0403MagnesiumRichFoods.pdf.

67 B. R. Stephens, K. Granados, T. W. Zderic, et al., "Effects of 1 Day of Inactivity on Insulin Action in Healthy Men and Women: Interaction with Energy Intake," *Metabolism* 60 (2011): 941–49.

68 M. S. Lunde, V. T. Hjellset, and A. T. Hostmark, "Slow Post Meal Walking Reduces the Blood Glucose Response: An Exploratory Study in Female Pakistani Immigrants," *Journal of Immigrant and Minority Health* 14 (2012): 816–22.

69 H. Nygaard, S. E. Tomten, and A. T. Hostmark, "Slow Postmeal Walking Reduces Postprandial Glycemia in Middle-aged Women," *Applied Physiology, Nutrition, and Metabolism* 34 (2009): 1087–92.

70 J. Henson, M. J. Davies, D. H. Bodicoat, et al., "Breaking up Prolonged Sitting with Standing or Walking Attenuates the Postprandial Metabolic Response in Postmenopausal Women: A Randomized Acute Study," *Diabetes Care* 39 (2016): 130–38.

71 P. C. Dempsey, R. N. Larsen, P. Sethi, et al., "Benefits for Type 2 Diabetes of Interrupting Prolonged Sitting with Brief Bouts of Light Walking or Simple Resistance Activities," *Diabetes Care* 39 (2016): 964–72.

72 S. F. Chastin, T. Egerton, C. Leask, et al., "Meta-analysis of the Relationship between Breaks in Sedentary Behavior and Cardiometabolic Health," *Obesity* 23 (2015): 1800–10.

73 E. Chacko, "Exercising Tactically for Taming Postmeal Glucose Surges," *Scientifica (Cairo)* (2016): 4045717.
74 R. J. Manders, J. W. van Dijk, and L. J. van Loon, "Low-intensity Exercise Reduces the Prevalence of Hyperglycemia in Type 2 Diabetes," *Medicine & Science in Sports & Exercise* 42, no. 2 (Feb. 2010): 219–25.
75 J. B. Gillen, B. J. Martin, M. J. MacInnis, L. E. Skelly, M. A. Tarnopolsky, and M. J. Gibala, "Twelve Weeks of Sprint Interval Training Improves Indices of Cardiometabolic Health Similar to Traditional Endurance Training despite a Five-fold Lower Exercise Volume and Time Commitment," *PLoS ONE* 11, no. 4 (2016): e0154075.
76 C. Benedict et al., "Gut Microbiota and Glucometabolic Alterations in Response to Recurrent Partial Sleep Deprivation in Normal-weight Young Individuals," *Molecular Metabolism* (published online Oct. 24, 2016), http://dx.doi.org/10.1016/j.molmet.2016.10.003.
77 A. J. Graveling, I. J. Deary, and B. M. Frier, "Acute Hypoglycemia Impairs Executive Cognitive Function in Adults with and without Type 1 Diabetes," *Diabetes Care* 36, no. 10 (Oct. 2013): 3240–46.
78 K. A. Page, A. Williamson, N. Yu, E. C. McNay, J. Dzuira, R. J. McCrimmon, and R. S. Sherwin, "Medium-Chain Fatty Acids Improve Cognitive Function in Intensively Treated Type 1 Diabetic Patients and Support in vitro Synaptic Transmission during Acute Hypoglycemia," *Diabetes* 58, no. 5 (May 2009): 1237–44.
79 Y. Nonaka, T. Takagi, M. Inai, S. Nishimura, S. Urashima, K. Honda, T. Aoyama, and S. Terada, "Lauric Acid Stimulates Ketone Body Production in the KT-5 Astrocyte Cell Line," *Journal of Oleo Science* 65, no. 8 (Aug. 2016): 693–99.
80 I. Hu Yang, J. E. De la Rubia Ortí, P. Selvi Sabater, S. Sancho Castillo, M. J. Rochina, N. Manresa Ramón, and I. Montoya-Castilla, "Coconut Oil: Non-alternative Drug Treatment against Alzheimer's Disease," *Nutrición Hospitalaria* 32, no. 6 (Dec. 2015): 2822–27.
81 A. J. Murray, N. S. Knight, M. A. Cole, L. E. Cochlin, E. Carter, K. Tchabanenko, T. Pichulik, M. K. Gulston, H. J. Atherton, M. A. Schroeder, R. M. Deacon, Y. Kashiwaya, M. T. King, R. Pawlosky, J. N. Rawlins, D. J. Tyler, J. L. Griffin, J. Robertson, R. L. Veech, and K. Clarke, "Novel Ketone Diet Enhances Physical and Cognitive Performance," *FASEB Journal* 30, no. 12 (Dec. 2016): 4021–32.

82 M. Vijayakumar, D. M. Vasudevan, K. R. Sundaram, S. Krishnan, K. Vaidyanathan, S. Nandakumar, R. Chandrasekhar, and N. Mathew, "A Randomized Study of Coconut Oil versus Sunflower Oil on Cardiovascular Risk Factors in Patients with Stable Coronary Heart Disease," *Indian Heart Journal* 68, no. 4 (Jul.–Aug. 2016): 498–506.

83 D. A. Cardoso, A. S. Moreira, G. M. de Oliveira, R. Raggio Luiz, and G. Rosa, "A Coconut Extra Virgin Oil-Rich Diet Increases HDL Cholesterol and Decreases Waist Circumference and Body Mass in Coronary Artery Disease Patients," *Nutricion Hospitalaria* 32, no. 5 (Nov. 2015): 2144–52.

84 J. K. Kiecolt-Glaser, M. A. Belury, R. Andridge, W. B. Malarkey, and R. Glaser, "Omega-3 Supplementation Lowers Inflammation and Anxiety in Medical Students: A Randomized Controlled Trial," *Brain, Behavior, and Immunity* 25, no. 8 (Nov. 2011): 1725–34.

85 R. Narendran, W. G. Frankle, N. S. Mason, M. F. Muldoon, and B. Moghaddam, "Improved Working Memory but No Effect on Striatal Vesicular Monoamine Transporter Type 2 after Omega-3 Polyunsaturated Fatty Acid Supplementation," ed. B. Le Foll, *PLoS ONE* 7, no. 10 (2012): e46832.

86 J. Bradbury, S. P. Myers, and C. Oliver, "An Adaptogenic Role for Omega-3 Fatty Acids in Stress; a Randomised Placebo Controlled Double Blind Intervention Study," *Nutrition Journal* 3 (2004): 20.

87 L. D. Lawson and B. G. Hughes, "Absorption of Eicosapentaenoic Acid and Docosahexaenoic Acid from Fish Oil Triacylglycerols or Fish Oil Ethyl Esters Co-ingested with a High-fat Meal," *Biochemical and Biophysical Research Communications* 156, no. 2 (Oct. 1988): 960–63.

88 X. Liu and T. Osawa, "Astaxanthin Protects Neuronal Cells against Oxidative Damage and Is a Potent Candidate for Brain Food," *Forum of Nutrition* 61 (2009): 129–35.

89 F. Shahidi and Y. Zhong, "Lipid Oxidation and Improving the Oxidative Stability," *Chemical Society Reviews* 39, no. 11 (Nov. 2010): 4067–79.

90 B. B. Albert, J. G. Derraik, D. Cameron-Smith, P. L. Hofman, S. Tumanov, S. G. Villas-Boas, M. L. Garg, and W. S. Cutfield, "Fish Oil Supplements in New Zealand Are Highly Oxidised and Do Not Meet Label Content of n-3 PUFA," *Scientific Reports* 5 (Jan. 2015): 7928.

91 M. Maes, R. Smith, A. Christophe, E. Vandoolaeghe, A. van Gastel, H. Neels, P. Demedts, A. Wauters, and H. Y. Meltzer, "Lower Serum

High-density Lipoprotein Cholesterol (HDL-C) in Major Depression and in Depressed Men with Serious Suicidal Attempts: Relationship with Immune-inflammatory Markers," *Acta Psychiatrica Scandinavica* 95, no. 3 (Mar. 1997): 212–21.

92 M. F. Muldoon, S. B. Manuck, and K. A. Matthews, "Lowering Cholesterol Concentrations and Mortality: A Quantitative Review of Primary Prevention Trials," *British Medical Journal* 301, no. 6747 (Aug. 1990): 309–14.

93 A. Gabriel, "Changes in Plasma Cholesterol in Mood Disorder Patients: Does Treatment Make a Difference?," *Journal of Affective Disorders* 99, nos. 1–3 (Apr. 2007): 273–78.

94 A. S. Wells, N. W. Read, J. D. Laugharne, and N. S. Ahluwalia, "Alterations in Mood after Changing to a Low-fat Diet," *British Journal of Nutrition* 79, no. 1 (Jan. 1998): 23–30.

95 L. Velázquez-López, A. V. Muñoz-Torres, C. García-Peña, M. López-Alarcón, S. Islas-Andrade, and J. Escobedo-de la Peña, "Fiber in Diet Is Associated with Improvement of Glycated Hemoglobin and Lipid Profile in Mexican Patients with Type 2 Diabetes," *Journal of Diabetes Research* 2016 (2016): 2980406.

96 H. E. Anderson-Vasquez, P. Pérez-Martínez, P. Ortega Fernández, and C. Wanden-Berghe, "Impact of the Consumption of a Rich Diet in Butter and it [*sic*] Replacement for a Rich Diet in Extra Virgin Olive Oil on Anthropometric, Metabolic and Lipid Profile in Postmenopausal Women," *Nutrición Hospitalaria* 31, no. 6 (Jun. 2015): 2561–70.

97 C. L. Kien, J. Y. Bunn, R. Stevens, J. Bain, O. Ikayeva, K. Crain, T. R. Koves, and D. M. Muoio, "Dietary Intake of Palmitate and Oleate Has Broad Impact on Systemic and Tissue Lipid Profiles in Humans," *American Journal of Clinical Nutrition* 99, no. 3 (Mar. 2014): 436–45.

98 Cardoso et al., "A Coconut Extra Virgin Oil-Rich Diet Increases HDL Cholesterol and Decreases Waist Circumference and Body Mass in Coronary Artery Disease Patients," 2144–52.

99 R. P. Mensink, P. L. Zock, A. D. Kester, and M. B. Katan, "Effects of Dietary Fatty Acids and Carbohydrates on the Ratio of Serum Total to HDL Cholesterol and on Serum Lipids and Apolipoproteins: A Meta-analysis of 60 Controlled Trials," *American Journal of Clinical Nutrition* 77, no. 5 (May 2003): 1146–55.

100 E. H. Temme, R. P. Mensink, and G. Hornstra, "Comparison of the Effects of Diets Enriched in Lauric, Palmitic, or Oleic Acids on Serum

Lipids and Lipoproteins in Healthy Women and Men," *American Journal of Clinical Nutrition* 63, no. 6 (Jun. 1996): 897–903.

101 N. B. Cater, H. J. Heller, and M. A. Denke, "Comparison of the Effects of Medium-chain Triacylglycerols, Palm Oil, and High Oleic Acid Sunflower Oil on Plasma Triacylglycerol Fatty Acids and Lipid and Lipoprotein Concentrations in Humans," *American Journal of Clinical Nutrition* 65, no. 1 (Jan. 1997): 41–45.

102 C. Stough, A. Scholey, J. Lloyd, J. Spong, S. Myers, and L. A. Downey, "The Effect of 90 day Administration of a High Dose Vitamin B-complex on Work Stress," *Human Psychopharmacology* 26, no. 7 (Oct. 2011): 470–76.

103 A. Pipingas, D. A. Camfield, C. Stough, A. B. Scholey, K. H. Cox, D. White, J. Sarris, A. Sali, and H. Macpherson, "Effects of Multivitamin, Mineral and Herbal Supplement on Cognition in Younger Adults and the Contribution of B Group Vitamins," *Human Psychopharmacology* 29, no. 1 (Jan. 2014): 73–82.

104 O. P. Almeida, A. H. Ford, V. Hirani, V. Singh, F. M. van Bockxmeer, K. McCaul, and L. Flicker, "B Vitamins to Enhance Treatment Response to Antidepressants in Middle-aged and Older Adults: Results from the B-VITAGE Randomised, Double-blind, Placebo-controlled Trial," *British Journal of Psychiatry* 205, no. 6 (Dec. 2014): 450-57.

105 O. P. Almeida, K. Marsh, H. Alfonso, L. Flicker, T. M. Davis, and G. J. Hankey, "B-vitamins Reduce the Long-term Risk of Depression after Stroke: The VITATOPS-DEP Trial," *Annals of Neurology* 68, no. 4 (Oct. 2010): 503–10.

106 D. O. Kennedy, R. Veasey, A. Watson, F. Dodd, E. Jones, S. Maggini, and C. F. Haskell, "Effects of High-dose B Vitamin Complex with Vitamin C and Minerals on Subjective Mood and Performance in Healthy Males," *Psychopharmacology* 211, no. 1 (Jul. 2010): 55–68.

107 O. Stanger, B. Fowler, K. Piertzik, M. Huemer, E. Haschke-Becher, A. Semmler, et al., "Homocysteine, Folate and Vitamin B12 in Neuropsychiatric Diseases: Review and Treatment Recommendations," *Expert Review of Neurotherapeutics* 9 (2009): 1393–1412.

108 K. Yoshino, M. Nishide, T. Sankai, M. Inagawa, K. Yokota, Y. Moriyama, et al., "Validity of Brief Food Frequency Questionnaire for Estimation of Dietary Intakes of Folate, Vitamins B6 and B12 and Their Associations with Plasma Homocysteine Concentrations," *International Journal of Food Sciences and Nutrition* 61 (2010): 61–67.

109 A. Oulhaj, F. Jernerén, H. Refsum, A. D. Smith, and C. A. de Jager, "Omega-3 Fatty Acid Status Enhances the Prevention of Cognitive Decline by B Vitamins in Mild Cognitive Impairment," *Journal of Alzheimer's Disease* 50, no. 2 (Jan. 2016): 547–57.

110 M. Montava, S. Garcia, J. Mancini, Y. Jammes, J. Courageot, J. P. Lavieille, and F. Feron, "Vitamin D3 Potentiates Myelination and Recovery after Facial Nerve Injury," *European Archives of Oto-Rhino-Laryngology* 272, no. 10 (Oct. 2014): 2815–23.

111 J. F. Chabas, D. Stephan, T. Marqueste, S. Garcia, M. N. Lavaut, C. Nguyen, R. Legre, M. Khrestchatisky, P. Decherchi, and F. Feron, "Cholecalciferol (Vitamin D3) Improves Myelination and Recovery after Nerve Injury," *PLoS ONE* 8, no. 5 (May 2013): e65034.

112 O. Józefowicz, J. Rabe-Jabłońska, A. Woźniacka, and D. Strzelecki, "Analysis of Vitamin D Status in Major Depression," *Journal of Psychiatric Practice* 20, no. 5 (Sept. 2014): 329–37.

113 C. Grudet, J. Malm, A. Westrin, and L. Brundin, "Suicidal Patients Are Deficient in Vitamin D, Associated with a Pro-inflammatory Status in the Blood," *Psychoneuroendocrinology* 50C (Sept. 2014): 210–19.

114 R. Ramadan, V. Vaccarino, F. Esteves, D. S. Sheps, J. D. Bremner, P. Raggi, and A. A. Quyyumi, "Association of Vitamin D Status with Mental Stress-induced Myocardial Ischemia in Patients with Coronary Artery Disease," *Psychosomatic Medicine* 76, no. 7 (Sept. 2014): 569–75.

115 A. Gholamrezaei, Z. S. Bonakdar, L. Mirbagher, and N. Hosseini, "Sleep Disorders in Systemic Lupus Erythematosus. Does Vitamin D Play a Role?" *Lupus* 23, no. 10 (Sept. 2014): 1054–58.

116 V. K. Wilson, D. K. Houston, L. Kilpatrick, J. Lovato, K. Yaffe, J. A. Cauley, T. B. Harris, E. M. Simonsick, H. N. Ayonayon, S. B. Kritchevsky, and K. M. Sink, "Health, Aging and Body Composition Study. Relationship between 25-Hydroxyvitamin D and Cognitive Function in Older Adults: The Health, Aging and Body Composition Study," *Journal of the American Geriatrics Society* 62, no. 4 (Apr. 2014): 636–41.

117 A. L. Peterson, C. Murchison, C. Zabetian, J. B. Leverenz, G. S. Watson, T. Montine, N. Carney, G. L. Bowman, K. Edwards, and J. F. Quinn, "Memory, Mood, and Vitamin D in Persons with Parkinson's Disease," *Journal of Parkinson's Disease* 3, no. 4 (2013): 547–55.

118 D. E. Bredesen, "Reversal of Cognitive Decline: A Novel Therapeutic Program," *Aging* 6, no. 9 (2014): 707–17.

119 L. Matsuoka, L. Ide, J. Wortsman, J. MacLaughlin, and M. F. Holick, "Sunscreens Suppress Cutaneous Vitamin D3 Synthesis," *Journal of Clinical Endocrinology & Metabolism* 64 (1987): 1165–68.

CHAPTER 8

1   K. S. Kendler, J. M. Hettema, F. Butera, C. O. Gardner, and C. A. Prescott, "Life Event Dimensions of Loss, Humiliation, Entrapment, and Danger in the Prediction of Onsets of Major Depression and Generalized Anxiety," *Archives of General Psychiatry* 60 (2003): 789–96.

2   F. Grabenhorst and E. T. Rolls, "Value, Pleasure and Choice in the Ventral Prefrontal Cortex," *Trends in Cognitive Sciences* 15 (2011): 56–67.

3   M. G. Craske, A. E. Meuret, T. Ritz, M. Treanor, and H. J. Dour, "Treatment for Anhedonia: A Neuroscience Driven Approach," *Depression and Anxiety* 33, no. 10 (Oct. 2016): 927–38.

4   S. Dalm, E. R. de Kloet, and M. S. Oitzl, "Post-training Reward Partially Restores Chronic Stress Induced Effects in Mice," ed. G. Chapouthier, *PLoS ONE* 7, no. 6 (2012): e39033.

5   N. Geschwind, F. Peeters, N. Jacobs, P. Delespaul, C. Derom, E. Thiery, J. van Os, and M. Wichers, "Meeting Risk with Resilience: High Daily Life Reward Experience Preserves Mental Health," *Acta Psychiatrica Scandinavica* 122, no. 2 (Aug. 2010): 129–38.

6   G. S. Alexopoulos and P. Arean, "A Model for Streamlining Psychotherapy in the RdoC Era: The Example of 'Engage,'" *Molecular Psychiatry* 19, no. 1 (Jan. 2014): 14–19.

7   D. Becker and J. van der Pligt, "Forcing Your Luck: Goal-striving Behavior in Chance Situations," *Motivation and Emotion* 40 (2016): 203–11.

8   M. Lehne and S. Koelsch, "Toward a General Psychological Model of Tension and Suspense," *Frontiers in Psychology* 6 (2015): 79.

9   S. H. Kim, Y. H. Kim, and H. J. Kim, "Laughter and Stress Relief in Cancer Patients: A Pilot Study," *Evidence-Based Complementary and Alternative Medicine* 2015 (2015): 864739.

10   K. Hayashi, I. Kawachi, T. Ohira, K. Kondo, K. Shirai, and N. Kondo, "Laughter is the Best Medicine? A Cross-Sectional Study of

Cardiovascular Disease Among Older Japanese Adults." *Journal of Epidemiology* 26, no. 10 (Oct. 2016): 546–52.

11 A. J. Blood, R. J. Zatorre, P. Bermudez, and A. C. Evans, "Emotional Responses to Pleasant and Unpleasant Music Correlate with Activity in Paralimbic Brain Regions," *Nature Neuroscience* 2 (1999): 382–87.

12 H. G. MacDougall and S. T. Moore, "Marching to the Beat of the Same Drummer: The Spontaneous Tempo of Human Locomotion," *Journal of Applied Physiology* 99 (2005): 1164–73.

13 D. Moelants, "Preferred Tempo Reconsidered," in *Proceedings of the 7th International Conference on Music Perception and Cognition*, eds. C. Stevens, D. Burnham, G. McPherson, E. Schubert, and J. Renwick (Adelaide, South Australia: Causal Production, 2002), 580–83.

14 C. J. Bacon, T. R. Myers, and C. I. Karageorghis, "Effect of Music-movement Synchrony on Exercise Oxygen Consumption," *Journal of Sports Medicine and Physical Fitness* 52 (2012): 359–65.

15 C. I. Karageorghis, P. C. Terry, A. M. Lane, D. T. Bishop, and D. L. Priest, "The BASES Expert Statement on Use of Music in Exercise," *Journal of Sports Sciences* 30, no. 9 (May 2012): 953–56.

16 N. Guegen and C. Jacob, "The Influence of Music on Temporal Perceptions in an On-hold Waiting Situation," *Psychology of Music* 30 (2002): 210–14.

17 J. Gibbon, R. M. Church, and W. H. Meck, "Scalar Timing in Memory," *Annals of the New York Academy of Sciences* 423 (1984): 52–77.

18 S. Droit-Volet, D. Ramos, J. L. Bueno, and E. Bigand, "Music, Emotion, and Time Perception: The Influence of Subjective Emotional Valence and Arousal?" *Frontiers in Psychology* 4 (Jul. 2013): 417.

19 A. Nguyen, L. Frobert, I. McCluskey, P. Golay, C. Bonsack, and J. Favrod, "Development of the Positive Emotions Program for Schizophrenia: An Intervention to Improve Pleasure and Motivation in Schizophrenia," *Frontiers in Psychiatry* 7 (Feb. 2016): 13.

20 G. F. Koob, "Addiction is a Reward Deficit and Stress Surfeit Disorder," *Frontiers in Psychiatry* 4 (Aug. 2013): 72.

21 T. S. Shippenberg, A. Zapata, and V. I. Chefer, "Dynorphin and the Pathophysiology of Drug Addiction," *Pharmacology & Therapeutics* 116, no. 2 (Nov. 2007): 306–21.

CHAPTER 9

1 G. L. Shulman, J. A. Fiez, M. Corbetta, R. L. Buckner, F. M. Miezin, et al., "Common Blood Flow Changes across Visual Tasks: II.

Decreases in Cerebral Cortex," *Journal of Cognitive Neuroscience* 9 (1997): 648–63.

2   M. E. Raichle, A. M. MacLeod, A. Z. Snyder, W. J. Powers, D. A. Gusnard, and G. L. Shulman, "A Default Mode of Brain Function," *PNAS* 98 (2001): 676–82.

3   R. L. Carhart-Harris and K. J. Friston, "The Default-mode, Ego-functions and Free-energy: A Neurobiological Account of Freudian Ideas," *Brain* 133, no. 4 (2010): 1265–83.

4   M. Kent, C. T. Rivers, and G. Wrenn, "Goal-directed Resilience in Training (GRIT): A Biopsychosocial Model of Self-regulation, Executive Functions, and Personal Growth (Eudaimonia) in Evocative Contexts of PTSD, Obesity, and Chronic Pain," *Behavioral Sciences* 5, no. 2 (Jun. 2015): 264–304.

5   http://www.venchar.com/2005/01/the_stockdale_p.html.

6   A. Bandura, *Self-efficacy: The Exercise of Control* (New York: Freeman, 1997).

7   K. Simmen-Janevska, V. Brandstätter, and A. Maercker, "The Overlooked Relationship between Motivational Abilities and Posttraumatic Stress: A Review," *European Journal of Psychotraumatology* 3 (2012): doi: 10.3402/ejpt.v3i0.18560.

8   A. Diedrich, S. G. Hofmann, P. Cuijpers, and M. Berking, "Self-compassion Enhances the Efficacy of Explicit Cognitive Reappraisal as an Emotional Regulation Strategy in Individuals with Major Depressive Disorder," *Behaviour Research and Therapy* 82 (Jul. 2016): 1–10.

A FINAL NOTE ON RESILIENCE

1   B. M. Iacoviello and D. S. Charney, "Psychosocial Facets of Resilience: Implications for Preventing Posttrauma Psychopathology, Treating Trauma Survivors, and Enhancing Community Resilience," *European Journal of Psychotraumatology* 5 (Oct. 2014): 10.3402/ejpt.v5.23970.

Index

About the Author

As Mithu Storoni became a doctor, dabbled in neuroscientific research, trained in ophthalmic surgery, taught yoga, and earned a PhD in neuro-ophthalmology, she studied "illness" through different angles, all of which led to the brain. The deeper she ventured into the study of "illness" the more aware she became of the vital role played by chronic stress in the transformation from "wellness" to "illness." Moving from fast-paced London to the breakneck Asian business hub of Hong Kong heightened this awareness as Mithu watched her thriving friends and colleagues reluctantly abandon their passion, energy, and mental agility and surrender their flourishing careers, because of chronic, unbridled stress. This prompted Mithu to write this book.

Mithu received her medical degree from the University of Cambridge and conducted her research in neuro-ophthalmology at the National Hospital for Neurology and Neurosurgery in London, where she was a Clinical Research Fellow until recently moving to Hong Kong with her husband.